THE SUNDAY TIMES

Brain Teasers

Book 2

200 mind-boggling riddles

Published in 2020 by Times Books

HarperCollins*Publishers*
Westerhill Road
Bishopbriggs
Glasgow, G64 2QT

www.collinsdictionary.com

10 9 8 7 6 5 4 3 2 1

The *Sunday Times* is a registered trademark of Times Newspapers Ltd

ISBN 978-0-00-840415-4

Typeset by Davidson Publishing Solutions, Glasgow

Printed and bound by CPI Group (UK) Ltd, Croydon CR0 4YY

If you would like to comment on any aspect of this book, please contact us
at the above address or online via email: puzzles@harpercollins.co.uk

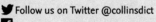 Follow us on Twitter @collinsdict
Facebook.com/CollinsDictionary

MIX
Paper from
responsible sources
FSC™ C007454

This book is produced from independently certified FSC™ paper
to ensure responsible forest management.

For more information visit: www.harpercollins.co.uk/green

CONTENTS

INTRODUCTION

Welcome to this new collection of Brain Teasers. Here we bring together 200 mind-bending puzzles from the series, which appears every week in *The Sunday Times* under the shortened name of Teaser.

Each Brain Teaser takes the form of a paragraph or two of text, sometimes accompanied by a diagram. The puzzles invariably require the application of mathematical or logical reasoning to reach the answer, and, like any true intellectual challenge, completing one offers great satisfaction to the solver.

The majority of the Brain Teasers in this book were originally published between October 2010 and October 2014. Thirteen are more recent, dating from the autumn of 2018. Two brilliant colleagues have overseen the development and publication of each Brain Teaser that appears in these pages. Dr Victor Bryant edited the puzzle for more than 40 years before handing the baton to John Owen at the end of 2017. I am immensely grateful to Victor and John for all their efforts in making the series such a success over their years at the helm.

Both editors also contribute puzzles to the series, complementing those submitted by many other regular and occasional setters. The seemingly endless ingenuity of our setters is nothing short of astonishing, and I would like to thank them all for the entertainment they continue to provide. The setter of each puzzle in this book is given at the back.

All Brain Teaser submissions are unsolicited and anyone with a good idea is welcome to send in a puzzle. If you have been inspired by the challenges in this book and would like to write a Brain Teaser puzzle for publication in *The Sunday Times*, please email: puzzles.feedback@sunday-times.co.uk for more information.

David Parfitt
Puzzles Editor, *The Times* & *The Sunday Times*

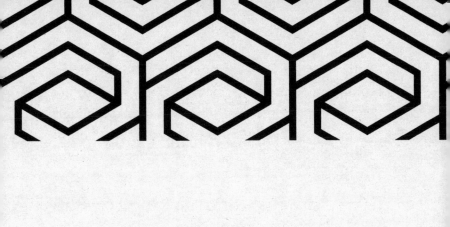

BRAIN
TEASERS

1 | IN HOT PURSUIT

George and Martha are jogging around a circular track. Martha starts at the most westerly point, George starts at the most southerly point, and they both jog clockwise at their own steady speeds.

After a short while Martha is due north of George for the first time. Five minutes later she is due south of him for the first time. Then George catches up with her during their second laps at the most northeasterly point of the track.

What is Martha's speed (in degrees turned per minute)?

2 | A LOPSIDED NUMBER

In this letters-for-digits substitution puzzle, each letter consistently represents a different digit. In the display, each letter in the top row is the sum of the two letters directly below it:

P O S E R

F I D D L E

What number is LOPSIDED?

3 | FIFTIETH ANNIVERSARY

We have recently celebrated the 50th anniversary of the Teaser column. At the party for the setters they gave each letter of the alphabet a different number from 1 to 26 (eg, they made A=7). Appropriately, this was done in such a way that, for each setter present, the values of the letters of their surname added up to 50. Angela Newing was there (so N+E+W+I+N+G=50), as were Nick MacKinnon and Hugh Bradley. Only two of Graham Smithers, Danny Roth, Andrew Skidmore, John Owen and Victor Bryant could make it.

Which two?

4 | EASTERTIDE

In the following division sum I have consistently replaced digits with letters, with different letters used for different digits. All three numbers featured are even:

EASTER ÷ TIDE = EGG

What is the numerical value of my TEASER?

5 | ELEMENTARY

"Every number has some significance," said Holmes, referring to his monograph on the subject. "Then what do you make of this?" asked Watson, scribbling a seven-digit number on the desk diary. "Apart from the obvious fact that it is your old Indian army number," replied Holmes, "I see that it is the largest seven-digit number where the difference between it and its reverse is the cube of a factor of the number itself."

What was Watson's number?

6 | ODDS AND EVENS

I have taken two three-figure numbers and multiplied them together by long multiplication. Below are my workings, but with O marking the positions of the odd digits and E the even digits:

```
        E   E   E
        O   O   O
    ─────────────────
    E   E   E   E   E
        E   O   E   E
        O   O   O   E
    ─────────────────
O   E   O   O   E
```

What are the two three-figure numbers?

7 | OCTAHEDRA

Fabule's latest jewellery creation consists of a set of identically sized regular octahedra made of solid silver. On each octahedron, Fabule has gold-plated four of its eight equilateral triangle faces. No two octahedra are the same, but if Fabule had to make another, then it would be necessary to repeat one of the existing designs.

How many octahedra are there in the set?

8 | DOUBLING UP

One day at school I found myself in charge of two classes. Faced with 64 pupils and only 32 desks, I gave each child a different whole number and told them the highest number must share with the lowest, the second highest with the second lowest, etc. When they had sorted themselves into pairs, I got each child to multiply their own number by that of their desk-mate. They discovered all the products were the same. In fact I chose the original numbers so that this common product would be as low as possible.

What was the common product?

9 | EXTRA TIME

George and Martha planned a holiday on the south coast. The second-class rail fare each way was a certain whole number of pounds per person and the nightly cost of an inexpensive double room, in pounds, was the same number but with digits reversed. They originally planned to stay 30 nights, but then increased that to 33. "So the total cost will go up by 10%," said Martha. "No," replied George, "it will go up by some other whole number percentage."

What is the nightly cost of a double room?

10 | LOTTO LUCK

Chris was lucky in the lotto, winning a whole number of pounds (under £5,000), which he shared with his five children. John got £1 more than a fifth of the total; Ciaran got £1 more than a fifth of the remainder. After Ciaran got his share, Fergal got £1 more than a fifth of the remainder, and after Fergal got his share, Brendan got £1 more than a fifth of what was left. After Brendan got his share, Chris gave Grainne £1 more than a fifth of what was left and kept the remainder (a whole number of pounds).

How much did Chris keep?

11 | TWO ROUTES

A line of equally spaced poles is marked in order with odd numbers, 1, 3, 5, etc. Directly opposite is a parallel line of an equal number of poles with even numbers, 2, 4, 6, etc. There are fewer than 100 poles. The distance in metres between adjacent poles is a two-digit prime; the distance between opposite poles is another two-digit prime.

Jan walks the route 1-2-4-3-5, etc to reach the final pole. John walks the odds 1-3-5, etc to the last odd-numbered pole; then walks diagonally to pole 2; then walks the evens 2-4-6, etc, also finishing at the final pole. Jan's and John's routes are of equal length.

How many poles are there?

12 | HIT AND MISS

For calculations using arithmetic in a base higher than 10, I need symbols for the higher "digits". My digits are 0, 1, 2, 3, 4, 5, 6, 7, 8, 9, A(=10), B(=11), C(=12), etc, using as many letters as necessary. But some numbers are ambiguous. Eg, in HIT and in MISS it is unclear whether the second digit is the number 1 or the letter I. So there are four interpretations of HIT + MISS.

I've taken one of those four interpretations and worked out its remainder when dividing by four. If you knew the remainder you could work out which base I am using.

What is that base?

13 | NEIGHBOURLY NONPRIMES

I live on a long road with houses numbered 1 to 150 on one side. My house is in a group of consecutively numbered houses where the numbers are all nonprime, but at each end of the group the next house number beyond is prime. There are a nonprime number of houses in this group.

If I told you the lowest prime number which is a factor of at least one of my two next-door neighbours' house numbers, then you should be able to work out my house number.

What is it?

14 | EMBLEMATIC

Pat's latest art installation consists of a large triangle with a red border and, within it, a triangle with a green border. To construct the green triangle, he drew three lines from the vertices of the red triangle to points one third of the way (clockwise) along their respective opposite red sides. Parts of these lines formed the sides of the green triangle.

In square centimetres, the area of the red triangle is a three-digit number and the area of the green triangle is the product of those three digits.

What is the area of the red triangle?

15 | SECRET AGENTS

Secret agents James and George exchange coded messages. The code is given by an addition sum, with different letters replacing different digits. The message is then written below the sum. James needs to send George the time (24-hour clock) at which their next secret operation is to begin. He texts as follows:

THREE + THREE + TWO = EIGHT
time: DATA

When does this secret operation begin?

16 | MULTIPLE CELEBRATION

Today is my birthday and the birthday of my granddaughter Imogen. My age today is a whole-number multiple of hers, and this has been true on one third of our joint anniversaries.

If we both live until I am six times her current age, then my age will be a multiple of hers on two more birthdays.

How old are we today?

17 | PLANETARY LINE

George and Martha are studying a far-off galaxy consisting of 10 planets circling a sun clockwise, all in the same plane. The planets are Unimus (taking one year to circle the sun), Dubious (two years), Tertius (three years), and so on up to Decimus (10 years).

At one instant today the sun and all 10 planets were in one straight line. Martha said it will be ages before that happens again. "Don't let's be greedy," said George. "How long must we wait until at least nine planets and the sun are in a straight line?"

How long must they wait?

18 | A ROUND TRIP

I own a circular field with six trees on its perimeter. One day I started at one tree and walked straight to the next, continuing in this way around the perimeter from tree to tree until I returned to my starting point. In this way I found that the distances between consecutive trees were 12, 12, 19, 19, 33 and 33 metres.

What is the diameter of the field?

19 | CRICKET AVERAGES

Playing for our local team, Sam and Oliver between them took 5 wickets in each innings, taking 5 wickets each overall. Sam's averages (ie, runs per wicket) were lower than Oliver's in both innings, but overall Sam had the higher average. All six averages were single non-zero digits.

If you knew Sam's overall average, it would then be possible to calculate the number of runs scored against Sam and against Oliver.

What were the total runs scored against (a) Sam and (b) Oliver?

20 | TAKE NOTHING FOR GRANTED

In arithmetic, the zero has some delightful properties. For example:

ANY + NONE = SAME

and

X . ZERO = NONE

In that sum and product, digits have been replaced with letters, different letters being used for different digits. But nothing should be taken for granted: here the equal signs are only approximations as, in each case, the two sides may be equal or differ by one.

Please send in your NAME

21 | EVER-INCREASING CIRCLES

I took a piece of A4 paper and cut it straight across to make a square and a rectangle. I then cut the square along a diagonal and, from the rectangle, cut as large a circle as I could. My friend Ron said that if I needed two triangles of those sizes and as large a circle in one piece as possible from an A4 sheet, then he could do much better. With a fresh sheet of A4, he produced the two triangles and the biggest circle possible.

What is the area of his circle divided by the area of mine?

22 | THE X-FACTOR

George and Martha's daughter entered a singing competition. The entrants were numbered from one upwards. The total number of entrants was a three-figure number and their daughter's number was a two-figure number. The five digits used in those two numbers were all different and non-zero. Martha noted that the number of entrants was a single-digit multiple of their daughter's number. George realised the same would be true if the two digits of their daughter's number were reversed.

How many entrants were there?

23 | ODD AND EVEN

Roddy and Eve have a set of cards on which is written a consecutive sequence of whole numbers, one number on each card. Roddy challenges Oliver to choose two cards at random, promising a prize if the sum of the numbers is odd. Eve issues the same challenge, but offers a prize for an even sum. Oliver accepts the challenge that gives him the greater probability of winning. This probability is a whole number percentage and, when reversed, it gives the two-figure number of cards.

Whose challenge did Oliver accept and how many cards are there?

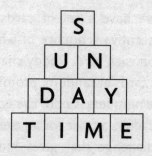

These 10 children's bricks are numbered from 1 to 10. Where a brick rests on two others, its number is the difference of their two numbers. Given that U=1...

What is the number DAY?

25 | DOTTY SQUARES

On a piece of paper I have drawn a neat, evenly spaced square array of 36 dots, namely six rows of six.

If I were to ask you how many squares are formed, you might say just 25, for the array seems to consist of 25 little squares. However, lots of sets of four of the dots form the vertices of a square.

How many ways are there of choosing four of the dots so that they form the vertices of a square?

26 | THE LEGACY

In my will I have set aside a five-figure sum of pounds for charity – a group of four animal charities and a group of seven children's charities. This five-figure sum consists of five consecutive non-zero digits, not in numerical order. On my death this legacy is to be divided equally between these charities. But I have left it to my executors' discretion to choose to divide it instead equally among the charities in just one of the groups. Each donation will be a whole number of pounds, whichever course they take.

What is the five-figure sum?

27 | ROUND TABLE

Pat likes to demonstrate his favourite trick shot on the pub pool table, which is 90in long and 48in wide. Pat hits the ball so that it strikes the long side of the table first, then (after a perfect bounce) it hits the short side, then the other long side, and then finally the other short side, from which it returns to the starting point, making its path the perimeter of a quadrilateral.

What is the length of the perimeter of that quadrilateral?

28 | INCONSEQUENTIAL

Most whole numbers can be expressed as the sum of two or more consecutive whole numbers. For example, 35=17+18, 36=11+12+13, and 40=6+7+8+9+10.

I have written down two whole numbers. Between them they use each of the digits 0 to 9 exactly once. But neither of the numbers can be expressed as the sum of two or more consecutive whole numbers.

What are my two numbers?

29 | THREE TREES

Our front garden is circular, with a diameter less than 100 metres. Three trees grow on the perimeter: an ash, a beech and a cherry, each a different whole number of metres from each of the other two. A straight trellis 84 metres long joins the ash to the beech, and a honeysuckle grows at the point on the trellis nearest to the cherry tree. The distance from the cherry to the ash plus the distance from the ash to the honeysuckle equals the distance from the cherry to the beech.

What is the diameter of the garden?

30 | WINNING THE DOUBLE

Four teams, A, B, C and D, played each other once in a league, then played a knockout competition. They scored a total of 54 goals in the nine games, with a different score in each game; no team scored more than five goals in any game. For each team the average number of goals per game was a different whole number. Team B beat D by a margin of more than one goal in the knockout competition, but team A took the double, winning all their games.

What was the score in the match CvD?

31 | CHEZ NOUS

A sign-writer stocks signs of house numbers written in words, from "one" up to my own house number. The signs are kept in boxes according to the number of letters used. Eg, all copies of "seven" and "fifty" are in the same box. Each box contains at least two different signs. Boxes are labelled showing the number of letters used – eg, the box holding "fifty" is labelled "five".

Naturally, the sign-writer has used signs from his own stock for the labels. To label all the boxes, he needed to take signs from half of the boxes.

What is my house number?

32 | PRIME TIME

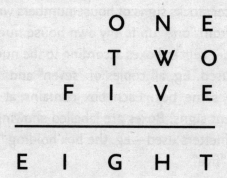

In this addition sum I have consistently replaced digits by letters, with different letters for different digits, and I have ended up with another correct addition sum.

What is the prime number NOW?

33 | 11/11/11

As it was 11/11/11 last Friday, I have been looking at numbers that are divisible by 11.

I have written down an eight-figure number using the digits 2, 3, 4, 5, 6, 7, 8 and 9 in some order. For any two adjacent digits in the number, either one is a multiple of the other or the two digits differ by one.

My number is divisible by its first digit, its last digit, and (of course) by 11.

What is the eight-figure number?

34 | RED AND BLACK

I have taken 10 cards and written a different digit on each one. Some of the cards are red and the rest are black.

I have placed some of the red cards in a row to form a long number. Then I have moved the last card to the front to give me a larger number. In fact this larger number divided by the original one equals a digit on one of the black cards.

What was the original number?

35 | A FULFILLING STRATEGY

I drove down a road with a number of petrol stations whose locations I saw on my map. I decided to check the price at the first station then fill up when I found one offering a lower price (or, failing that, the last one).

Looking back, I see each price per litre was different. Overall they were consecutive whole numbers of pence plus 0.9p in each case (ie, 130.9p, 131.9p, 132.9p, etc, not necessarily in that order). So I'm surprised to discover the average price was a whole number of pence per litre.

How many petrol stations were there?

36 | A CLOSE THING

Our football league consists of five teams, each playing each other once, earning three points for a win, one for a draw. Last season was close. At the end, all five teams had tied on points and on "goal difference".

So the championship was decided on "goals scored", which, for the five teams, consisted of five consecutive numbers. The scores in the games were all different, including a no-score draw, and no team scored more than five goals in any game.

How many goals did the league champions score in total?

37 | ENDING 2011

I have written down three positive whole numbers consisting of two primes and a perfect square. Between them they use nine digits and no digit is repeated.

If you form an addition sum with the three numbers, then what do you end up with as the total? Appropriately enough you end up with 2011.

What, in increasing order, are the three numbers?

38 | CHRISTMAS GIFT

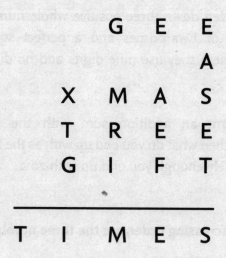

```
      G E E       E
              A
    X   M A S
    T   R E E       E
    G   I F T
    _____

  T I M E S
```

I sent Hank, an American friend, a Times subscription for a Christmas present. He is pleased and has responded with this numerical addition sum with a prime total in which he has consistently replaced digits by letters, with different letters for different digits.

What is TIMES?

39 | CHRISTMAS TEASER

I asked my nine-year-old grandson Sam to set a Teaser for today's special edition and the result was:

SAM SET NICE CHRISTMAS TEASER

ThosE words are the result of taking five odd multiples of nine and consistently replacing digits by letters. Given that THREE is divisible by 3:

What is the nine-digit number CHRISTMAS?

40 | DOG SHOW

Five dogs were entered by owners Alan, Brian, Colin, David and Edward, whose surnames are Allen, Bryan, Collins, Davis and Edwards. No owner had the same first and second initials, no two had the same pair of initials and none shared an initial with their breed. Colin was in position 1, David Allen's dog was in position 4, Brian's corgi was not next to the collie; the chow and doberman were at opposite ends. The other breed was a dachshund. In the voting, dogs eliminated in order were Mr Bryan's, the corgi, David's and the doberman.

Who won, and what breed was their dog?

41 | ABOMINABLE

In the art class Pat used five circles for his new design, Snowman. One circle made the head and another touching circle (with radius three times as big) made the body. Two equal circles (of radius one centimetre less than that of the head) made the arms (one on each side) and they touched both the body and the head. The fifth circle enclosed the other four, touching each of them. The radius of the head's circle was a whole number of centimetres.

How many centimetres?

42 | PRIME CRICKETERS

George and Martha support their local cricket team and keep records of all the results. At the end of last season, they calculated the total number of runs scored by each of the 11 players. George commented that the totals were all different three-figure palindromic primes. Martha added that the average of the 11 totals was also a palindromic prime.

What were the two highest individual totals?

43 | HIGHER POWERS

I have chosen two different two-figure numbers. They use the same two digits as each other, but in different orders. If I start with either number and raise it to a power equal to either of the two numbers, then I find that the answer is a number whose last two digits form the number with which I started.

What are my two numbers?

44 | MORE TEAMS

In our local pub soccer league each team plays each of the others at home and away during the season. Two seasons ago we had a two-figure number of teams. The number increased the next year, and again this year, but this time the increase was one less than the previous season. This season the number of games played will be over three times the number played two seasons ago. Also, this season's increase in games played actually equals the number played two seasons ago.

How many teams are there now in the league?

45 | LATIN CUBE

I have a cube and on each face of it there is a capital letter. I look at one face from the front, hold the top and bottom faces horizontal, and rotate the cube. In this way I can correctly read a four-letter Roman numeral. Secondly, I hold the cube with the original front face underneath and the original top face at the front, and I rotate the cube keeping the new top and bottom faces horizontal. In this way I correctly see another four-letter Roman numeral. Thirdly, I return the cube to its original position and rotate it with the two sides vertical to give another correct four-letter Roman numeral. The sum of the second and third numbers is less than a hundred.

What was the third four-letter Roman numeral seen?

46 | NUMBER WAVES

I challenged Sam to find a nine-figure number using all of the digits 1 to 9, such that each digit occupying an even position was equal to the sum of its adjacent digits. (For example, the digit occupying position four would be equal to the sum of the digits occupying positions three and five.) Sam produced a solution that was exactly divisible by its first digit, by its last digit, and by 11.

What was his number?

47 | THE RIGHT MONEY

In order to conserve his change, Andrew, our local shopkeeper, likes to be given the exact money. So today I went to Andrew's shop with a collection of coins totalling a whole number of pounds in value. I could use these coins to pay exactly any amount from one penny up to the total value of the coins. Furthermore, the number of coins equalled the number of pounds that all the coins were worth; I had chosen the minimum number of coins to have this property.

(a) How many coins did I have? (b) For which denominations did I have more than one coin of that value?

48 | BOX CLEVER

I have placed the digits 1 to 9 in a three-by-three grid made up of nine boxes. For each two-by-two array within the grid, the sum of its four entries is the same. Furthermore, if you reverse the order of the digits in that common total, then you get the sum of the four corner entries from the original grid.

Which digits are adjacent to the 6? (ie, whose boxes have a common side with 6's box?)

49 | HARD TIMES

	1	2	3	4	...
1	1	2	3	4	...
2	2	4	6	8	...
3	3	6	9	12	...
4	4	8	12	16	...
⋮	⋮	⋮	⋮	⋮	⋮

Penny found a multiplication table in the back of one of Joe's old school books – the top corner of it is illustrated. She noticed that prime numbers only appeared twice in the body of the table, whereas 4 (for example) appeared three times and 12 appeared six times. Penny could not help wondering how many times large numbers appeared in a huge table.

What is the smallest number that will appear 250 times?

50 | RESPLENDENCY

Our local Hibernian band will lead the St Patrick's Day parade next Saturday, resplendent in their new tunics and finery. The total cost of this was one thousand pounds and it was generously paid for by the parish council. Each tunic cost the same two-figure number of pounds (and that number happens to be double the number of girls in the band). In addition, each girl in the band has some extra finery, and the cost of this for each girl was the same as the cost of the tunic but with the two digits in reverse order.

How many boys and how many girls are there in the band?

51 | ANYONE FOR TENNIS?

The girls of St Trinian's choose two games from the four on offer. In Felicity's gang of four, no game was chosen by both Daphne and Erica; Chloe and Miss Brown chose lacrosse; Miss Smith chose netball; and only Miss Jones excluded hockey. In Harriet's gang of four, Clara, Debbie and Ellen all chose netball but only one of the four chose hockey. To avoid having two girls with the same first name initial in any one game, one of the girls in the second group (but not Debbie) moved from one game to another. This meant that there was an odd number of these girls playing each game.

Which of the eight girls played tennis?

52 | PLAYING CARDS

Oliver wrote down some numbers and then used a letter-for-digit code to make them into words. In this way his numbers became JACK, QUEEN, KING and ACE. In fact JACK and KING were perfect squares, QUEEN was odd, and ACE was a prime. Furthermore, appropriately enough, the sum of the four was divisible by 52.

What number was represented by QUEEN?

53 | TRUNCATION

Callum is learning about squares and primes. He told me that he had discovered a four-figure square that becomes a three-figure prime if you delete its last digit. I could not work out his square from this information and he informed me that, if I knew the sum of its digits, then I would be able to work it out. So I then asked whether the square had a repeated digit: his answer enabled me to work out his square.

What was it?

54 | EASTER TEASER

I took the letters A, E, E, R, S and T and wrote down the long list of all possible different combinations of them. So, in alphabetical order, my list was AEERST, AEERTS, ..., EASTER, ..., TEASER, ..., TSREEA. I then assigned each of the letters a different odd digit, turning my list into a list of numbers. Surprisingly, the grand total of my list of numbers contained no odd digit.

What is E+A+S+T+E+R?

55 | POWERFUL DICE

George and Martha threw a dice, noted the number of spots on the uppermost face, and entered that number in one of the boxes of a three-by-three grid of nine squares. They repeated the exercise eight more times resulting in a number in each box.

Then George looked at the three three-figure numbers read across the rows and commented that each was a square or a cube. Martha looked at the three three-figure numbers read down the columns and said the same was true for her numbers. Furthermore, their two sets of three numbers had only one in common.

What (in increasing order) were the five different three-figure numbers?

56 | HOBSON'S CHOICE

A	B	C	D
E	F	G	H
I	J	K	L
M	N	O	P

In the diagram the letters represent the numbers 1 to 16 in some order. They form a magic square, where the sum of each row, each column and each main diagonal is the same. The letters of the word "CHOICE" and some others (including all the letters not used in the magic square) have a value equal to their position in the alphabet (C=3, H=8 etc).

What is the value of H+O+B+S+O+N?

57 | ON-LINE

Linesmen Alf and Bert were checking the railway line from Timesborough to Teaserton: Alf started at Timesborough, Bert at Teaserton, and they walked towards each other at their own steady speeds. A train from Timesborough, travelling at constant speed, reached Alf at noon and passed him ten seconds later. At 12.06 the train reached Bert and passed him eight seconds later.

At what time did the two men meet?

58 | A CERTAIN AGE

My cousin and I share a birthday. At the moment my age, in years, is the same as hers but with the order of the two digits reversed. On one of our past birthdays I was five times as old as her but, even if I live to a world record "ripe old age", there is only one birthday in the future on which my age will be a multiple of hers.

How old am I?

59 | MOVING DOWNWARDS

A three-by-three array of the nine digits 1 – 9 is said to be "downward moving" if each digit is less than the digit to the east of it and to the south of it – see, for example, the keypad on a mobile phone.

Megan has produced a downward moving array that is different from that on a mobile phone. If she were to tell you the sum of the digits in its left-hand column and whether or not the central digit was even, then you should be able to work out Megan's array.

What is Megan's array?

60 | SHILLY CHALET

George and Martha are running a holiday camp and their four daughters are staying there. To keep the peace they have been given chalets in different areas. Amelia's chalet number is between 1 and 99, Bertha's is between 100 and 199, Caroline's between 200 and 299, and Dorothy's between 300 and 399.

George commented that the difference between any two of the four chalet numbers is either a square or a cube. Martha added that the same could be said for the sum of the chalet numbers of the three youngest daughters.

Who is the eldest daughter and what is her chalet number?

61 | INVENTIVE INVENTORIES

It is known that there is just one ten-figure number that is "self-counting" in the sense that its first digit equals the number of 1s in it, the second digit equals the number of 2s in it, and so on, with the ninth digit equalling the number of 9s in it and the tenth digit equalling the number of 0s in it.

Similarly, a nine-figure number is "self-counting" if its first digit equals the number of 1s in it, the second digit equals the number of 2s in it, and so on, with the ninth digit equalling the number of 9s in it (and with the 0s remaining uncounted).

(a) What is the ten-figure self-counting number?
(b) What is the highest nine-figure self-counting number?

62 | SPELL IT OUT!

I have written down a very large number (but with fewer than twenty-five digits). If you spell out each of its digits as a word, then for each digit its last letter is the same as the first letter of the next digit (the last letter of the last digit being the same as the first letter of the first digit). One example of such a number would be 83821. Neither my number nor the sum of its digits is a palindromic number (but in fact the original number is one more than a palindrome).

What is my number?

63 | JUST ENOUGH

I have two identical bags of balls, each containing two reds, two yellows and two blues. Blindfolded throughout, I remove just enough balls from the first to be sure that the removed balls include at least two colours. I place these balls in the second bag. Then I remove some balls from the second bag and put them in the first: I move just enough to be sure that the first bag will now contain at least one ball of each colour.

What (as a percentage) is the probability that all the balls left in the second bag are the same colour?

64 | STYLISH FENCES

Farmer Giles has a hexagonal field bounded by six straight fences. The six corners lie on a circle of diameter somewhere around 250 metres. At three alternate corners there are stiles. The angles of the hexagon at the corners without stiles are all the same. All six fences are different whole numbers of metres long. I have just walked in a straight line from one of the stiles to another and the distance walked was a whole number of metres.

How many?

65 | UNUSUAL FACTORS

There are no neat tests for divisibility by 7 or 13, so it's unusual for these factors to feature in Teasers: today we put that right. If you start with any three different non-zero digits, then you can make six different three-figure numbers using precisely those digits. For example, with 2, 5 and 7 you get 257, 275, 527, 572, 725 and 752. Here, one of their differences (572 – 257) is divisible by 7, and another (725 – 257) is divisible by 13.

Your task today is to find three different digits so that none of the six possible three-figure numbers and none of the differences between any pair of them is a multiple of either 7 or 13.

What are the three digits?

66 | AGES AGO

Bert's age in years is one less than one-and-a-half times the age Alf was a whole number of years ago. Cal's age in years is one less than one-and-a-half times the age Bert was, the same number of years ago. Dave's age in years is one less than one-and-a-half times the age Cal was, again the same number of years ago. All four ages are different two-figure numbers, Cal's age being Bert's age with the order of the digits reversed.

What (in alphabetical order) are their ages?

67 | DIE HARD

I have placed three identical standard dice in a row on a table-top, leaving just eleven faces and an odd number of spots visible. Regarding each face as a digit (with, naturally enough, the five-spotted face being the digit 5, etc) from the front I can read a three-figure number along the vertical faces and another three-figure number along the top. Similarly, if I go behind the row I can read a three-figure number along the vertical faces and another three-figure number along the top. Of these four three-figure numbers, two are primes and two are different perfect squares.

What are the two squares?

68 | REMOVING A CARD

I have taken a set of cards and written a different digit on each. Overall I have used a set of consecutive digits. Then I have arranged the cards in some order to make one large number. This number is divisible by the digit that is one more than the number of cards. Reading from left to right, if any card is removed then the remaining number is divisible by the position of the card removed. For example, if the card in the sixth position is removed and the remaining cards are pushed together to form another number, then that number is divisible by 6.

What was the original number?

69 | MILEOMETER

My car's digital mileometer has five digits, displaying whole miles. It also has a trip meter with five digits, displaying tenths of a mile (which resets to zero when reaching 10,000 miles). Soon after the car was new I reset the trip meter to zero (but never did that again). Not long after, I noticed that the two five-digit displays were the same (ignoring the decimal point). The next time the displays were the same I wrote down the mileage and did so on each subsequent occasion until the mileage reached 100,000. The sum of all the mileages I wrote down was 500,000.

What were the five digits when the displays were first the same?

70 | OLYMPIC ECONOMICS

George and Martha are organising a large sports meeting. Usually a gold medal (costing £36) is given to each winner, silver (£17) to each runner-up, and bronze (£11) to each third. In a minority of events, such as tennis and boxing, both losing semi-finalists get a bronze medal. However, George and Martha are going to economise by having third place play-offs in such events, thus reducing the medals needed by an odd number. George noticed that the total cost of the medals is now a four-figure palindrome, and Martha commented that the same would have been true under the previous system.

How many events are there?

71 | OVER FIFTY YEARS

We have now had over fifty years of Teasers and to celebrate this I found three special numbers using only non-zero digits and I wrote the numbers down in increasing order. Then I consistently replaced all the digits by letters, with different letters for different digits. In this way the numbers became

FIFTY YEARS TEASER.

In fact the third number was the sum of the other two.

Which number does TEASER represent?

72 | PANDIGITAL SQUARE

I call my telephone number "pandigital", in the sense that it is a nine-figure number using each of the digits 1 to 9. Amazingly, it is a perfect square. Furthermore, its square root is a five-figure number consisting of five consecutive digits in some order. It might interest you to know (although you do not need to) that my neighbour's telephone number is also a nine-figure pandigital perfect square, but his is at least double mine. With a little logic and not many calculations you should be able to work out my telephone number.

What is it?

73 | FOREIGN FRIENDS

The telephone number of my Japanese friend is a ten-figure number written as a group of five two-figure numbers. The number does not start with a 0, no digit is used more than once, the numbers in the group are in increasing order, no two of them have a common factor, and one of them is a fourth power. Also, the product of the numbers in the group is divisible by four of the first five primes.

The telephone number of my Russian friend is a ten-figure number but it is written as a group of two five-figure numbers. The number and group have the same properties as those stated above. Its second digit is double the second digit of the Japanese number.

What are the two telephone numbers?

74 | SIMPLE SUMS

I have taken some numbers and consistently replaced the digits by letters, with different letters for different digits. In this way:

TWO + TWO = FOUR
FIVE IS PRIME
FIVE - TWO IS PRIME
THREE IS DIVISIBLE BY 3

Quelle est la valeur de HUIT?

75 | CUE FOR A QUEUE

Alan, Brian, Colin, Dave and Ed have surnames Smith, Jones, Rogers, Mason and Hall, but not in that order. They form themselves into a queue. Brian is somewhere ahead of Mr Smith who is somewhere ahead of Ed. Similarly, Mr Jones is ahead of Colin who is ahead of Dave who is ahead of Mr Hall. Mr Mason's two neighbours in the queue are Alan and Mr Rogers.

If I told you Alan's surname it would now be possible to work our all their surnames and their positions in the queue.

What is Alan's surname and what is his position in the queue?

76 | TRIBONACCI

In the Fibonacci sequence 1, 2, 3, 5, 8, 13, ... each number after the second is the sum of the previous two. Pat has experimented with some "Tribonacci" sequences of positive whole numbers where each number after the third is the sum of the previous three. He chooses the first three numbers to be different and in increasing order and then generates the sequence. He has shown us one where just the first five numbers are less than a hundred and where a later number is ten thousand.

What are the first three numbers in this sequence?

77 | FOOTBALL LOGIC

In the logicians' football tournament, each of three teams (captained by Alf, Bert and Charlie) plays each other once, with three points for a win and one for a draw. In working out their order at the end of the tournament, "goals scored" are used to determine the order of teams with the same points total. Each captain only knows the scores in his own team's games. At the end I asked the captains in turn whether they knew which position they had finished in. The replies were: Alf "no"; Bert "no"; Charlie "no"; Alf "yes".

In which position did Charlie's team finish, and what was the result in the game between the other two teams?

78 | STRING ART

Callum knocked some tacks onto the edge of a circular board. He then used pieces of string to make all possible straight-line connections between pairs of tacks. The tacks were arranged so that no three pieces of string crossed at the same point inside the circle. If he had used six tacks this would have required fifteen pieces of string and it would have created fifteen crossing points inside the circle. But he used more than six and in his case the number of crossing points inside the circle was a single-digit multiple of the number of pieces of string used.

How many tacks did he use?

79 | ODD AGES

Recently Alex, Bernard and Charles were comparing their ages in years. Alex's age was an odd two-figure number and Bernard's age was a whole number percentage more than that. On the other hand, Charles's age was also an odd two-figure number but Bernard's age was a whole number percentage less than that. They were surprised to find that the percentage was the same in both cases.

What were their three ages?

80 | SIMPLE ARITHMETIC

George and Martha are teaching their great-grandchildren some simple arithmetic. "If you add two thirties to four tens you get a hundred," George was saying, and he wrote it like this:

```
    T H I R T Y
    T H I R T Y
        T E N
        T E N
        T E N
        T E N
H U N D R E D
```

"Strangely," added Martha, there are nine different letters there, and if you allow each letter to stand for a different digit, there is a unique sum which works."

Which digit would be missing?

81 | STAR SIGNS

Five friends, Abel, Bob, Jedd, Tim and Will, were investigating birthdays. Abel commented that there was just one letter of the alphabet that occurred in both his star sign and in his month of birth. After further investigation he found that for any pair of his star sign, his month of birth, his name, and the day of the week on which he was born, just one letter of the alphabet occurred in both. Remarkably, the same turned out to be true for each of the friends.

Their names are listed alphabetically. What, respectively, are their star signs?

82 | ANAGRAMMATICAL

The number 258 is special in the following way. If you write down its six different anagrams (258, 285, 528, etc), then calculate the fifteen differences between any two of those six (27, 270, 243, etc) and then add up those fifteen differences, you get 4644, which is a multiple of the number 258 that you started with!

Which higher three-figure number has the same property?

83 | THE END OF TIME

A	B	C	D	E
F	G	H	I	J
K	L	M	N	O
P	Q	R	S	T
U	V	W	X	Y

In the table the letters represent the numbers 1 to 25 in some order. In each row, each column and each main diagonal the sum of the five numbers is the same. If you list all the letters in increasing numerical order then somewhere in the list you will get ... , S, U, N, D, A, Y, T, I, M, ... with E coming later.

Which letters are equal to 11, 9, 7, 3, 23 and 22?

84 | GUNPOWDER, TREASON AND PLOT

Last year I overheard this conversation one day during the first seven days of November: Alec: "It's the 5th today, let's have a bonfire." Bill: "No, the 5th is tomorrow." Chris: "You're both wrong – the 5th is the day after tomorrow." Dave: "All three of you are wrong." Eric: "Yesterday was certainly not the 1st." Frank: "We've missed bonfire night." If you knew how many of their statements were true, then you could work out the date in November on which this conversation took place.

What was that date?

85 | ELEVENSES

On 11/11 it is perhaps appropriate to recall the following story. In the graphics class students were illustrating pairs of similar triangles. In Pat's larger triangle the sides were all even whole numbers divisible by 11. In fact they were 22 times the corresponding sides of his smaller triangle. As well as this, in the smaller triangle the digits used overall in the lengths of the three sides were all different and did not include a zero. Miraculously, exactly the same was true of the larger triangle.

What were the sides of Pat's smaller triangle?

86 | A CASE OF AMBIGUITY

Every schoolchild knows that it is possible for two triangles to have two of their lengths-of-sides in common and one of their angles in common without the two triangles being identical or "congruent". This can happen when the common angle is not included between the two common sides. I have found such a pair of non-congruent triangles with all the lengths of the sides being a whole number of centimetres and with the longest side of each triangle being 10cm in length.

What are the lengths of the other two sides of the smaller triangle?

87 | ANTICS ON THE BLOCK

A wooden cuboid-shaped block has a square base. All its sides are whole numbers of centimetres long, with the height being the shortest dimension. Crawling along three edges, an ant moves from a top corner of the block to the furthest-away bottom corner. In so doing it travels ten centimetres further than if it had chosen the shortest route over the surface of the block.

What are the dimensions of the block?

88 | FIRST AND LAST

A "First and last" number (FLN) is any number that is divisible by its first and last digits. Using each non-zero digit just once, I wrote down a three-figure FLN followed by three two-figure FLNs. Putting these in one long line in that order gave me a nine-figure FLN. When the first and last digits were removed from this nine-figure number, I was left with a seven-figure FLN.

What was the nine-figure number?

89 | STATING THE OBVIOUS

I have replaced the letters of the alphabet by the numbers 0 to 25 in some order, using different numbers for different letters. So, for example, a three-letter word could represent a number of three, four, five or six digits. With my particular values ONE + ONE = TWO, and ONE is a perfect square.

What number does TWO represent?

90 | COME DANCING

Angie, Bianca, Cindy and Dot were given dance partners and performed in front of four judges, Juno, Kraig, Len and Marcia. Each judge placed the performances in order and gave 4 marks to the best, then 3, 2 and 1 point to the others. The dancers' marks were then added up and they finished in alphabetical order with no ties.

Angie's winning margin over Bianca was larger than Bianca's over Cindy, and Bianca's winning margin over Cindy was larger than Cindy's over Dot. Juno's ordering of the four was different from the final order, and Kraig's 4 points and Len's 3 points went to dancers who were not in the top two.

How many points did Cindy get from Juno, Kraig, Len and Marcia respectively?

91 | CHRISTMAS MESSAGE

In the message below the words represent numbers where the digits have consistently been replaced by letters, with different letters used for different digits.

A MERRY XMAS 2012 TO YOU

In fact the largest of these six numbers is the sum of the other five.

What number is represented by MERRY?

92 | LATECOMER

Imogen is having a New Year party tomorrow night and she has invited Madeline, hoping not to have a repeat of last year's episode which completely spoilt the celebrations. Last New Year's Eve Madeline had been due at a certain whole number of minutes past an hour but she was late, the number of minutes late being that same aforementioned "certain number of minutes" but with the digits in the reverse order. Madeline pointed out that at least the angle between the hands of the clock at the moment of her arrival was the same as it would have been when she was due.

At what time was she due to arrive?

93 | IS IT 2013?

I have in mind a four-figure number. Here are three clues about it:

- Like 2013, it uses four consecutive digits in some order.
- It is not divisible by any number from 2 to 11.
- If I were to tell you the _____ digit, then you could work out my number.

Those clues would have enabled you to work out my number but unfortunately the printer has left a gap: it should have been the word "units" or "tens" or "hundreds" or "thousands".

What is my number?

94 | MIND THE GAP

If you list in increasing order all ten-figure numbers with no repeated digit, then the first is 1023456789 and the last is 9876543210. The differences between consecutive numbers in the list vary wildly but no difference is ever equal to the average of all the differences between the consecutive numbers.

What difference between consecutive numbers comes closest to the average?

95 | OUR SECRET

Wo, who lives in Woe, has been studying homophones; ie, sets of words that sound the same, but have different meanings. He has now extended his studies to sets of numbers that have a common property. He wrote down three numbers which were prime and a further two numbers whose different digits formed a consecutive set. Having consistently replaced different digits by different letters, this made his primes into

TWO TOO TO

and his numbers using consecutive digits into

STRAIGHT STRAIT.

What is the three-figure number WOE?

96 | INVERSION

In the woodwork class Pat cut a triangle from a sheet of plywood, leaving a hole in the sheet. The sides of the triangle were consecutive whole numbers of centimetres long. He then wanted to turn the triangle over and fit it back into the hole. To achieve this he had to cut the triangle into three pieces. He did this with two straight cuts, each cut starting at a midpoint of a side. Then each of the three pieces (two of which had the same perimeter) could be turned over and placed back in exactly its starting position in the hole.

What are the lengths of the sides of the original triangle?

97 | UNNATURALLY QUIET

I have given each letter of the alphabet a different value from 0 to 25, so some letters represent a single digit and some represent two digits. Therefore, for example, a three-letter word could represent a number of three, four, five or six digits. With my values it turns out that NATURAL = NUMBER.

What is the sum of the digits in the number MUTE?

98 | RANDOM ROAD

George and Martha moved into Random Road, which has 99 houses along just one side. But instead of being numbered 1 to 99 in the usual way, the man given the job of numbering them gave the first house a number equal to his age and then kept adding his age again for each subsequent house, ignoring the hundreds digits and above. (So, had he been 37 then the numbers would have been 37, 74, 11, 48, 85, 22, etc.) Luckily each house got a different number. George's house number was "correct" so he did not immediately notice. He only saw that something was amiss when Martha pointed out that a house next door had a number like theirs, but with the two digits in reverse order.

What is George's and Martha's house number, and how old is the numberer?

99 | RIVER CROSSING

Two lads walking together had to get back to their tent which was a short distance south-east of them. However, it involved crossing a river which was running from west to east and was a constant five metres wide. Whichever route they chose, they made sure that they were in the water for the shortest distance possible. One lad took the obvious such route and went due south and then due east, but the other took the shortest possible such route, thus cutting his friend's distance by a quarter.

What was the length of that shortest route?

100 | FAMILY PROGRESSION

I have five grandchildren: from oldest to youngest they are Imogen, Madeline, Alex, Matthew and Laurence. They were all born in different years and no two have a birthday in the same month. They were all born before noon and for each grandchild I have noted their time and date of birth as five numbers: minute, hour, DD, MM, YY (none of the numbers is zero, but the DD, MM or YY can have a leading zero).

Surprisingly, for each grandchild the five numbers form an arithmetic progression (ie, increasing or decreasing by a fixed amount), and also in each case the sum of two of the five numbers equals the sum of the other three.

With the children in the order given above, list the months in which they were born.

101 | TIMES TABLE

I have placed the numbers 1 to 16 (in no particular order) in a four-by-four table. As you go across each row of the table (from left to right) the four numbers are increasing, and as you go down each column the four numbers are increasing. If you read the top row as one long number, it is a four-figure number which equals the product of the four numbers in the second row.

What are the four numbers in the second row?

102 | MAGIC MUSHROOMS

Enid and her famous five (Anne, Dick, George, Julian and Timmy) were joined by Pixie in a mushroom hunt. They each picked two varieties, the fourteen overall being Bird's Nest, Beef Steak, Blue Toothed, Cannon Ball, Death Cap, Devil's Um, Inky Cap, Old Man of the Woods, Parasol, Poison Pie, Stinkhorn, Slippery Jack, Tree Ears and The Gypsy. For each of these hunters, if you wrote down their name and their two varieties of mushrooms, then for any two from the three you would find that there were just two letters of the alphabet that occurred in both.

What mushrooms were picked by (a) Pixie and (b) Enid?

103 | GOOD COMPANY

Each year at this time Pat's company gives its staff a bonus to help them "drown their shamrocks" on the Irish national holiday. A total of £500 is shared out amongst all six employees (five men and the manager Kate) whose numbers of years of service consist of six consecutive numbers. Each man gets the same whole number of pounds for each year of his service and Kate gets a higher whole number of pounds for each year of her service. This means that, although Pat does not get the lowest bonus, he does get £20 less than Kate – even though he has served the company for a year longer than her.

How much does Pat get?

104 | THREE SISTERS

In Shakespeare's less well-known prequel, King Lear shared all of his estate amongst his three daughters, each daughter getting a fraction of the estate. The three fractions, each in their simplest form, had numerators less than 100 and had the same denominators. Cordelia got the least, with Regan getting more, and Goneril the most (but her share was less than three times Cordelia's). Each of the three fractions gave a decimal which recurred after three places (as in 0.abcabc...) and each digit from 1 to 9 occurred in one of the three decimals.

What were the three fractions?

105 | SIMPLE EASTER TEASER

I have written down three numbers, at least one of which is odd. Furthermore, one of the three is the sum of the other two. Then I have consistently replaced digits by letters, using different letters for different digits, and the numbers have become

SIMPLE EASTER TEASER

What number is EASTER?

[Incidentally, Easter Sunday 2013 has a palindromic date, 31/3/13. Interestingly, this happens next in 2031.]

106 | POWERFUL TEAM

George and Martha support the local football team. The squad is numbered from 1 to 22, with 11 playing at any one time and the remaining 11 sitting on the bench. At the start of this week's match Martha commented that among their 11 players on the pitch there were no three consecutive numbers, and that the same was true of the 11 sitting on the bench. She also commented that the sum of the 11 players' numbers on the pitch was a perfect square. At half-time one substitution was made and in the second half George noted that all that Martha had said was still true, but now the perfect square was higher.

What (in increasing order) were the numbers of the 11 players in the first half?

107 | DIGILIST

I have written down a list of positive whole numbers in ascending order. Overall they use each of the digits 0 to 9 exactly once. There is more than one even number in the list, and the majority of the numbers are primes. The final number in the list is a perfect square and it is the sum of all the other numbers in the list.

What is my list of numbers?

108 | PRIME NUMBER

I have given each letter of the alphabet a different whole-number value between 1 and 99 so that each letter represents one or two digits. In this way, for example, a three-letter word can represent a number of three, four, five or six digits. With my values it turns out that

PRIME = NUMBER

Furthermore, rather fortuitously, each letter used in that display has a value equal to an odd prime number.

What is the number PRIME?

109 | MEGAN'S NUMBER

I had ten cards and each was labelled with a different digit. Megan chose three of these cards and arranged them to give a number. Beth then chose some of the remaining cards and also arranged them as a number. Finally Jan took some, or all, of the remaining cards and produced her number.

If Megan multiplied her number by its last digit, she would get Beth's number, and if Megan multiplied her number by its first digit she would get Jan's number.

Who has the biggest number and what is it?

110 | EXCHANGE RATE

I am going on holiday to Tezarland. So at our local Bureau de Change I exchanged a three-figure number of pounds for Tezar dollars. From the pounds I paid a £12 fee and received a whole number of Tezar dollars for each remaining pound. This was an unremarkable transaction except that it was an exchange in more ways than one: the number of Tezar dollars I received was like the number of pounds that I started with but with the digits in the reverse order.

How many Tezar dollars did I receive?

111 | NEW WORLD ORDER

In maths class Pat has been learning about progressions from his "New World" text book. The book has three sections, namely "Arithmetic", "Algebra" and "Geometry" which overall run from page 1 to page 500 inclusive – with each section starting on a new page. The geometry section is over twice as long as the arithmetic section.

As an exercise Pat calculated the sum of the page numbers in each section and he was surprised to find that the sum of the page numbers in "Algebra" was double that of the sum of the page numbers in "Geometry".

How many pages are there in the "Arithmetic" section?

112 | PACK POINTS

Five cubs were awarded points for effort. Enid's son and Master Smith had 17 points between them, Masters Jones and Robinson had a total of 16, Ivan and Master Robinson together had 14, and the two youngest of the cubs had a total of 13. John and Master Brown had a total of 12 points, Brenda's son and Mike had 11 points between them, Ken and his best friend had a total of 10, Ann's son and Debbie's son together had 9, Christine's son and Nigel had a total of 8, and Master Perkins and Debbie's son together had 6.

In alphabetical order of their mothers' Christian names, what were the names of the five cubs (Christian name and surname)?

113 | ROUTE CANAL

Stephen is planning a 70-mile canal trip from the lock at Aytown to the lock at Beeswick, stopping at a pub at a lock over half way along the route. He has listed the number of miles from each lock to the next, the largest being the stretch from the pub to the next lock. He has also noted the distances from Aytown to the pub and from the pub to Beeswick. All the numbers that he has written down are different primes. Stephen can use his figures to work out the number of miles from any lock to any other. He's found that, whenever that number of miles between locks is odd, then it is also a prime.

What (in order) are the numbers of miles between consecutive locks?

114 | ONE SQUARE

I have consistently replaced digits by letters, using different letters for different digits, and in this way I have found that

FIVE - FOUR SQUARED IS ONE.

Now if I were to tell you whether or not THREE is divisible by 3, and whether or not FOUR is divisible by 4, then you could work out FIVE.

What number is FIVE?

115 | MONDAY BIRTHDAYS

In one auspicious month last year our family had two great celebrations: Harry, the oldest member of the family, had his 100th birthday and, in the same month, his great-grandson Peter was born. It turns out that they were both born on the same day of the week. Of course, Harry has celebrated several of his 100 birthdays on a Monday, as will Peter. However, even if Peter lives that long, the number of his first 100 birthdays that fall on a Monday will be two fewer than Harry's.

On which day of the week were they born?

116 | HAPPY MEDIUM

At a recent séance Adam, Alan, Andy, Eden, Eric, Fred, Gary, Glen, John, Mike, Pete and Tony sat around a circular table of radius one metre. Around its edge there were 12 equally spaced points representing different letters of the alphabet.

A light started at the centre, moved straight to one of the points, moved straight to another, then straight to another, and so on, before returning directly to the centre. In this way it spelt out the name of one of the people present. It then started again and in a similar fashion spelt out a day of the week. Then it started again and spelt out a-month. Every straight line path that it took was a whole number of centimetres long.

Which three words did it spell out?

117 | PAINTED CUBES

Oliver found some painted cubes in the loft. These cubes had edges whose lengths were whole numbers of centimetres. After choosing some cubes whose edge lengths were consecutive, Oliver proceeded to cut them into "mini-cubes" of side one centimetre. Of course, some of these mini-cubes were partially painted and some were not painted at all. On counting up the mini-cubes, Oliver noticed that exactly half of them were not painted at all.

How many mini-cubes were not painted at all?

118 | RIGHT TO LEFT

I have given each letter of the alphabet a different value from 0 to 25, so some letters represent a single digit and some represent two digits. Therefore, for example, a three-letter word could represent a number of three, four, five or six digits. In fact the word CORRECT represents a nine-figure number. It turns out that

TO

REFLECT

RIGHTTOLEFT

are three even palindromes.

What is the CORRECT number?

119 | SQUARES OF OBLONGS

I gave to each of Ruby and Annie a set of "Oblongs". Each set consisted of nine pieces of card of sizes 1 2, 2 3, 3 4, 4 5, 5 6, 6 7, 7 8, 8 9 and 9 10. The idea was to use some or all of the cards to make various shapes in jigsaw fashion. I asked them to make a square using some of their own pieces. Ruby made the smallest square possible with her set and Annie made the largest square possible with hers. Then I collected up all the unused pieces of card.

What (in increasing order) were the sizes of these unused pieces?

120 | PARSNIP FARM

On Parsnip Farm there was a triangular field with one of its boundaries running north to south. From that boundary's southernmost point Farmer Giles built a fence in an easterly direction, thus partitioning the field into two smaller triangular fields, with the area of the northern field being forty per cent of the area of the southern field. Each side of each field was a whole number of furlongs in length, one of the sides of the southern field being twenty-five furlongs in length.

What were the lengths of the other two sides of the southern field?

121 | SQUARE MEALS

A company produces five different types of muesli. The ingredients are bought from a wholesaler who numbers his items from 1 to 10. In each type of muesli there is a mixture of a square number of different ingredients and the weight in grams of each ingredient is the square of its item number: also the total weight of its ingredients is a perfect square number of grams.

Last month one of the ingredients was unavailable and so only the "basic" and "fruity" varieties could be produced. This week a different ingredient is unavailable and so only the "luxury" variety can be made.

What are the item numbers of the ingredients in the luxury muesli?

122 | TOUGH GUYS

The SS Thomas More has two identical vertical masts mounted on the centre line of the deck, the masts being a whole number of feet tall and seven feet apart. Two straight guy ropes secure the top of each mast to a single anchor point on the centre line of the deck some distance forward of the masts. The total length of the two ropes is a whole number of feet, with one rope being two feet longer than the other.

What is the height of the masts?

123 | SQUARE CUT

Given a rectangular piece of paper that is not actually square it is possible with one straight cut to divide it into a square and a rectangle (which might or might not itself be square). I call this process a "square cut". Recently I started with a rectangle of paper one metre long with the width being more than half a metre. I performed a square cut and put the square to one side. On the remaining rectangle I performed a square cut and put the square to one side. I kept repeating this process until the remaining rectangle was itself a square. The result was that I had cut the original piece of paper into six squares all of whose sides were whole numbers of centimetres.

How many centimetres wide was the original piece of paper?

124 | SUDOPRIME

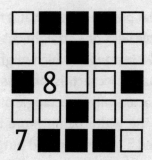

The grid shows a cross-figure with two of its digits given. The eleven answers (five of them being "across" and the other six "down") are all different prime numbers with no leading zeros. No digit appears more than once in any row, column or main diagonal.

What are the two largest of the eleven primes?

125 | WRONG ADDING

Here is an addition sum in which digits have consistently been replaced by letters, with different letters used for different digits. The six-figure total is even:

$$
\begin{array}{r}
\text{A G A I N} \\
\text{W R O N G} \\
\hline
\text{A D D I N G}
\end{array}
$$

Unfortunately, once again I have made a mistake. In just one place in the display I have written an incorrect letter.

What is the correct numerical value of the six-figure total?

126 | PUZZLING BOOK

George and Martha have a book of puzzles numbered from 1 to 30. The solutions are also numbered from 1 to 30, but a solution number is not necessarily the same as the number of the puzzle. George and Martha have solved some of the puzzles. If you look at the number of the puzzle and the number of the solution, then the sum of the two is a perfect power of the difference of the two. George has added up the numbers of the solved puzzles and got a three-figure total. Martha has added up the numbers of the solutions of the solved puzzles and got a higher three-figure total. In fact her total used the same non-zero digits as George's, but in a different order.

What (in increasing order) are the numbers of the solved puzzles?

127 | DIFFERENT VIEWS

Oliver arranged six die in a neat pile with three in the bottom row, two in the middle row and one at the top. The faces of these die were digits rather than the corresponding number of dots. Looking down on them, Beth saw that the five partially-visible tops of the die contained different digits. In the three rows at the front she saw one-figure, two-figure and three-figure primes, whereas from the back she saw three perfect squares. On the left, working down the three sides, she saw a three-figure square whereas on the right, again working down, she saw a three-figure prime.

What was this three-digit prime?

128 | TWO BY TWO

I have given each letter of the alphabet a different value from 0 to 25, so some letters represent a single digit and some represent two digits. Therefore, for example, a three-letter word could represent a number of three, four, five or six digits. With my values it turns out that

TWO X TWO = FOUR.

Another "obvious" fact that I can tell you is that every digit occurring in TWO is a prime!

What is the number FOUR?

129 | OBLONGING

I have made a set of "oblongs", which is what I call rectangles where the longer sides are one centimetre longer than the shorter sides. I made this set from some A4 sheets of card (without any pasting), cutting out one oblong of each of the sizes 1cm-by-2cm, 2cm-by-3cm, 3cm-by-4cm, and so on up to a particular size: I needed more than one A4 sheet to do this. I have given the set to my family and challenged them to use all the card in jig-saw fashion to make another oblong. After considerable effort they have managed to do this.

How many oblongs are in the set?

130 | THREE CLOCKS

I have three digital clocks, each showing the time as a four-digit display "hhmm" on the 24-hour scale. One keeps perfect time, one runs fast at a constant rate (less than a minute per day) and the third runs slow at exactly the same rate. Every six months I simultaneously reset the faulty clocks to the correct time. Recently I noticed that each clock displayed the same collection of digits but in different orders. In fact, on the fast clock no digit was in the correct place.

What time did the correct clock show at that moment?

131 | NUMBER PLEASE

I finally got through to the operator of a large company and asked to be connected to an appropriate department. He tried eight different extension numbers before then finding the correct one. The eight numbers were 1933, 2829, 3133, 4630, 5089, 5705, 6358 and 6542. Curiously, each of these wrong numbers did have at least one correct digit of the correct extension in the correct position!

What was the correct extension number?

132 | MISSED THE PLOT

My friend has a triangular vegetable plot, all sides being whole numbers of metres. Coincidentally, the dimensions of the plot are such that its perimeter (in metres) is the same as its area (in square metres). Also, the length of one of the sides is the average of the lengths of the other two sides.

What are the lengths of the sides of the plot?

133 | TIME 'N' AGAIN

Time and again at school we would be given the exercise of changing a fraction into a decimal. This time the given fraction is in its simplest form and it equals a recurring decimal. In some places the digits have been consistently replaced by letters, with different letters used for different digits, but in four places the digits have merely been replaced by asterisks.

$$\frac{\text{TIME}}{**\text{N}**} = .\text{AGAINAGAIN}...$$

Numerically, what is the TIME?

134 | MILKY WAYS

Each evening Pat drives his herd of cows into the milking parlour. The cows are ear-tagged 1,2,3, … and the parlour is divided into stalls also numbered 1,2,3, … , with one stall for each cow. The cows file in and choose empty stalls at random until all the stalls are full. Pat has noticed that very often it happens that no cow occupies a stall with the same number as her tag. Pat worked out the number of different ways this could happen, and also the number of ways that at least one cow could be in a stall with the same number as herself. The two answers were less than a million and Pat noticed that the sum of the digits in each was the same.

How many cows are in the herd?

135 | MULTIPLE CALLS

In my office each of the 100 employees has a different two-digit phone number, from 00 to 99. Recently my phone was rewired so that each number button generated an incorrect digit. Trying to phone four of my colleagues resulted in me calling double the intended number, and, for more than ten of my colleagues, trying to phone their number resulted in me calling triple the intended number. Also if I tried to phone any colleague and then asked for their phone number, then phoned that number and asked that person for their number, then phoned that number ... and so on, it always took ten calls to contact the person I intended.

If I tried to phone the numbers 01, 23, 45, 67 and 89 respectively, which numbers would I actually get?

136 | PRIME TIME

On a picture of a clock face I have written A next to one of the numbers. Then I counted a certain number of "hours" clockwise and wrote B. Then I continued in this pattern, always counting the same number of places clockwise and writing the next letter of the alphabet. In this way each letter corresponds to a number between 1 and 12. I noticed that the two numbers with three letters by them were prime. I also noticed that if I wrote the numbers corresponding to the letters of PRIME in order and read them as one long number, then I got a six-figure prime.

Which number corresponds to A and which to B?

137 | SMALL CUBES

Oliver and Megan each had a different-sized cuboid whose sides were all whole numbers of centimetres in length. Their cuboids were painted all over. They each cut their cuboid into one-centimetre cubes, some of which were unpainted, the rest being painted on one or more face. Oliver observed that for both cuboids, the number of cubes with no painted faces was the same as the number with two painted faces. Then Megan added "I have more cubes than you, and the difference between our totals is equal to the number of your unpainted cubes."

How many of Megan's cubes had just one painted face?

138 | EXTENDING THE GARDEN

George and Martha used to have a square garden with sides of length 10 metres, but they have extended it by adding an acute-angled triangle on each side of the square. The result is that the new garden has eight straight sides, each of which is fenced. The fences are different whole numbers of metres long, with the shortest being one metre and the longest 13 metres. George remarked that the average length of the fences was also a whole number of metres.

In increasing order, what were the lengths of the eight fences?

139 | ANSWERS ON A POSTCARD

On a postcard I have written four two-figure numbers, none of which is divisible by three. In three of the numbers the two digits used are consecutive (in some order) and in fact overall the four numbers use eight consecutive digits.

I have calculated the sum and product of the four numbers. Then I have consistently replaced each of the digits 0 to 9 by a different letter of the alphabet. It turns out that the sum of my four numbers is SUM and their product is PRODUCT.

What number is represented by POSTCARD?

140 | ON THE FACE OF IT

At the beginning of last year my wife and I were each given a clock. When we first looked at them the hands on both clocks showed the correct time but thereafter (although the two clocks coincided at least once more during the year) they generally showed different times. The problem was that my clock gained a certain whole number of minutes each day, and the same was true of my wife's clock, but hers was even faster than mine. Actually my clock showed the correct time at least once in every month last year but that was not the case for my wife's clock.

How many minutes did each clock gain in a day?

141 | 11, 12, 13 ...

On Wednesday the date will be 11.12.13. To celebrate this fact I have found three numbers with the lowest divisible by 11, another divisible by 12, and the remaining number divisible by 13. Furthermore, appropriately, the sum of the three numbers is 2013. Overall the three numbers use nine digits with no digit used more than once.

What (in increasing order) are my three numbers?

142 |FOOTPRINTS

A cubical dice, with faces labelled as usual, is placed in one of the nine squares of a three-by-three grid, where it fits exactly. It is then rotated about one of its edges into an adjacent square and this is done a total of eight times so that the dice visits each square once. The "footprint" of the route is the total of the nine faces that are in contact with the grid.

(a) What is the lowest footprint possible?
(b) What is the highest footprint possible?
(c) Which whole numbers between those two values cannot be footprints?

143 | ROLL MODEL

My new board game has squares numbered from 1 to 100 and has two unusual dice. The first die is ten-sided with numbers from 1 to 10, and the second is four-sided with a prime number on each side. A move consists of throwing the two dice and then choosing either one of the numbers or their sum and moving that number of squares in either direction. I found myself on one square and realised that there were just two squares which it would be impossible for me to reach with my next move. Both of those squares had prime numbers that did not appear on the dice.

(a) Which particular square was I on?
(b) What are the four numbers on the second die?

144 | SHOULD OLD ACQUAINTANCE ...

I have given each letter of the alphabet a different whole number value from 1 to 26 (eg, A=5 and B=24). In this way I can work out the value of any word by adding up the values of its letters. Looking at the values of the words

SHOULD OLD ACQUAINTANCE

I find that the value of OLD is the average of the other two values.

I also find that NEW and YEAR have the same value as each other – it is like the value of OLD but with its two digits in the reverse order. It is possible from all this information to work out the values of N, E and W, but all we want to know is ...

... what is the value of NEW?

145 | NEW DIARY

My diary has this design on the cover:

In this two-by-three grid there are 12 junctions (including the corners), some pairs of which are joined by a straight line in the grid. In fact there are 30 such pairs. The diary publisher has been using such grids of various sizes for years and the 1998 diary was special because its grid had precisely 1998 pairs of junctions joined by lines. Within the next decade they will once again be able to produce a special diary where the number of joined pairs equals the year.

(a) What was the grid size on the 1998 diary?
(b) What is the year of this next special diary?

146 | ONE FOR EACH DAY

George and Martha have been looking into tests for divisibility, including one for the elusive number seven. George wrote down a thousand-figure number by simply writing one particular non-zero digit a thousand times. Then he replaced the first and last digits by another non-zero digit to give him a thousand-figure number using just two different digits. Martha commented that the resulting number was divisible by seven. George added that it was actually divisible by exactly seven of 2, 3, 4, 5, 6, 7, 8, 9 and 11.

What were the first two digits of this number?

147 | MAP SNAP

I have two rectangular maps depicting the same area, the larger map being one metre from west to east and 75cm from north to south. I've turned the smaller map face down, turned it 90 degrees and placed it in the bottom corner of the larger map with the north-east corner of the smaller map touching the south-west corner of the larger map. I have placed a pin through both maps, a whole number of centimetres from the western edge of the larger map. This pin goes through the same geographical point on both maps. On the larger map 1cm represents 1km. On the smaller map 1cm represents a certain whole number of kilometres ...

... how many?

148 | PALPRIMES

My two pals and I have been considering "palprimes"; ie, palindromic numbers that are also prime. In particular each of us tried to find a five-figure palprime and I managed to come up with 39293. Then each of my two pals found a five-figure palprime. On comparing them we were surprised to find that overall our three palprimes used all the digits from 1 to 9.

What were the other two five-figure palprimes?

149 | YESTERDAY

In a new design of mobile phone each of the number buttons 1 to 9 is associated with two or three letters of the alphabet, but not in alphabetical order (and there are no letters on any other buttons). For example, M, T and U are on the same button. Predictive software chooses letters for you as you type. The numbers to type for SUNDAY, TIMES and TEASER are all multiples of 495.

What number should I type to make SATURDAY?

150 | INCONSEQUENTIAL

An "arithmetic" sequence is one in which each number is a fixed amount more than the previous one. So, for example, 10, 29, 48, ...is an arithmetic sequence. In this case its ninth number is 162, which happens to be divisible by 9. I have in mind another arithmetic sequence whose ninth number is divisible by 9. This time it starts with two three-figure numbers, but in this case I have consistently replaced digits by letters, with different letters used for different digits.

The arithmetic sequence then begins ONE, TWO, ... And its ninth number is NINE.

To win, please send in the number to WIN.

151 | 007

Eight villains Drax, Jaws, Krest, Largo, Morant, Moth, Sanguinette and Silva have been ordered to disrupt the orbits of the planets Earth, Jupiter, Mars, Mercury, Neptune, Saturn, Uranus and Venus, with each villain disrupting a different planet. Our agents Brosnan, Casenove, Connery, Craig, Dalton, Dench, Lazenby and Moore have each been assigned to thwart a different villain. For any villain/planet, villain/agent or planet/agent combinations just two different letters of the alphabet occur in both names.

List the eight agents in alphabetical order of their assigned villains.

152 | HOUSE-TO-HOUSE

Tezar Road has houses on one side only, numbered consecutively. The Allens, Browns, Carrs and Daws live in that order along the road, with the Allens' house number being in the teens. The number of houses between the Allens and the Browns is a multiple of the number of houses between the Browns and the Carrs. The number of houses between the Allens and the Carrs is the same multiple of the number of houses between the Carrs and the Daws. The Daws' house number consists of the same two digits as the Allens' house number but in the reverse order.

What are the house numbers of the four families?

153 | HOW SAFE?

Fed up with being burgled, George and Martha have bought a new safe. They remember its six-figure combination as three two-figure primes. Martha noted that the product of the six different digits of the combination was a perfect square, as was the sum of the six. George commented further that the six-figure combination was actually a multiple of the difference between that product and sum.

What is the six-figure combination?

154 | NOT THE GOLD CUP

Five horses took part in a race, but they all failed to finish, one falling at each of the first five fences. Dave (riding Egg Nog) lasted longer than Bill whose horse fell at the second fence; Big Gun fell at the third, and the jockey wearing mauve lasted the longest. Long Gone lasted longer than the horse ridden by the jockey in yellow, Chris's horse fell one fence later than the horse ridden by the jockey in green, but Fred and his friend (the jockey in blue) did not fall at adjacent fences. Nig Nag was ridden by Wally and Dragon's jockey wore red.

Who was the jockey in yellow, and which horse did he ride?

155 | GOOD INNINGS

Tomorrow is St Patrick's Day, so Pat will go on his usual tour of inns. Last year each inn had the same admission charge (made up of a two-figure number of pounds plus a two-figure number of pence surcharge). This four-figure total of pence happened to be the same as the four-figure number of pence Pat started the evening with, except that the order of the digits was reversed. In the first inn he spent one pound less than a third of what he had left after paying admission. In the second inn he also spent one pound less than a third of what he came in with after paying admission. That left him just enough to get into the third inn.

How much was that?

156 | CREASED

I took a rectangular piece of paper whose sides were whole numbers of centimetres – indeed the longer side was exactly three times the length of the shorter side. I folded the rectangle along a diagonal. I then took the folded piece and folded it again along its line of symmetry. Then I completely unfolded it to see that the two creases had divided the rectangle into two equal triangles and two equal quadrilaterals. In square centimetres the area of each triangle was a three-figure number, and the area of each quadrilateral was another three-figure number using the same three digits but in a different order.

What was the area of each triangle?

157 | MOTHER'S DAY

Today we are having a family get-together to celebrate Mother's Day. My maternal grandmother, my mother and I have each written down our date of birth in the form "ddmmyy". This gives us three six-figure numbers and, surprisingly, both of the ladies' numbers are multiples of mine. Furthermore, all of the digits from 0 to 9 occur somewhere in the three six-figure numbers.

What is my mother's six-figure date of birth?

I am organising a tombola for the fete. From a large sheet of card (identical on both sides) I have cut out a lot of triangles of equal area. All of their angles are whole numbers of degrees and no angle exceeds ninety degrees. I have included all possible triangles with those properties and no two of them are identical. At the tombola entrants will pick a triangle at random and they will win if their triangle has a right angle. The chances of winning turn out to be one in a certain whole number.

What is that whole number?

159 | LEAVING PRESENT

When maths teacher Adam Up reached the age of 65, he asked his colleagues for some spring bulbs as a leaving present. So they gave him some packs of bulbs with, appropriately enough, each pack containing 65 bulbs. Adam planted all the bulbs in clumps of different sizes, the number of bulbs in each clump being a prime number. Furthermore, these prime numbers overall used each of the ten digits exactly once. Had he been given any fewer packs this would have been impossible.

How many bulbs were there in the smallest and largest clumps?

160 | HOT X BUNS

I have consistently replaced each of the digits 0 to 9 by a different letter to turn some numbers into words. It turns out that the number EASTER is exactly divisible by each of

(a) the number of days in a week;
(b) the number of days in a year;
(c) the seasonal product HOTBUNS.

What numbers do HOT and BUNS represent?

161 | RUNNERS-UP

In our local football league each team plays each other team twice each season, once at home and once away. The teams get three points for a win and one point for a draw. All games are on Saturdays at 3pm, and yesterday all the teams played their last game, so the final league table has now been published. Our team were runners-up, but we should have won! We were just one point behind the winners and, although they won two-thirds of their games, our team won more. Furthermore, looking at our drawn games, I notice that the league winners lost twice as many games as we drew.

How many games did the league winners win, draw and lose?

162 | LET THE DOG SEE THE RABBIT

Each of Messrs Burrows, Cook, Field, Skinner and Warren has a pet dog, and each of the men's wives has a pet rabbit. Each dog bit one of the rabbits, with no two dogs biting the same rabbit. Mrs Warren's rabbit was bitten by Mr Skinner's dog. Mr Warren's dog bit the rabbit owned by the wife of the owner of the dog that bit Mrs Field's rabbit. Mrs Burrows' rabbit was bitten by the dog owned by the husband of the lady whose rabbit was bitten by the dog belonging to the husband of the lady whose rabbit was bitten by the dog belonging to Mr Cook.

Whose dog bit Mrs Skinner's rabbit, and whose rabbit did Mr Field's dog bite?

163 | IMPAIRED

Pat was given a rod of length 40 centimetres and he cut it into six pieces, each piece being a whole number of centimetres long, and with the longest piece twice as long as one of the other pieces. Then he took two of the pieces and wrote down their total length. He did this for every possible set of two pieces and he never got the same total more than once. However, when he started again and repeated the process but with sets of three pieces, one of the totals was repeated.

What (in increasing order) were the lengths of his pieces?

164 | POWERS BEHIND THE THRONES

Gold sovereigns were minted in London for most years from the great recoinage in 1817 until Britain left the gold standard in 1917. I have a collection of eight sovereigns from different years during that period, the most recent being an Edward VII sovereign (he reigned from 1902 until 1910). I noticed that the year on one of the coins is a perfect square and this set me thinking about other powers. Surprisingly, it turns out that the product of the eight years on my coins is a perfect cube.

What (in increasing order) are those eight years?

165 | DIGITAL DISPLAY

George and Martha have a seven-figure telephone number consisting of seven different digits in decreasing order. Martha commented that the seven digits added together gave a total number that did not use any of the digits from the telephone number. George agreed and added that their telephone number was in fact a multiple of that total number.

What is their telephone number?

166 | GIMME FIVE!

In the list of words below each different letter consistently represents a different digit. Each of the coded numbers has the property that the sum of its digits is divisible by (but not equal to) the number it spells out.

THREE FIVE EIGHT NINE TEN ELEVEN THIRTEEN

What number does FIVE represent?

167 | POND PLANTS

John bought some packs of pond plants consisting of oxygenating plants in packs of eight, floating plants in packs of four and lilies in packs of two, with each pack having the same price. He ended up with the same number of plants of each type. Then he sold some of these packs for twenty-five per cent more than they cost him. He was left with just fifty plants (with fewer lilies than any other) and he had recouped his outlay exactly.

How many of these fifty plants were lilies?

168 | LOW POWER

I have written down an addition sum and then consistently replaced digits by letters, with different letters used for different digits. The result is

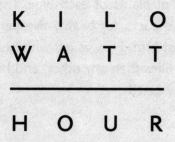

```
K I L O
W A T T
───────
H O U R
```

Appropriately enough, I can also tell you that WATT is a perfect power.

Please send in the three-figure LOW number.

169 | SPELL CHECK

This is Teaser 2700. The number 2700 is not particularly auspicious, but it does have one unusual property. Notice that if you write the number in words and you also express the number as a product of primes you get

TWO THOUSAND SEVEN HUNDRED = 2233355:

the number of letters on the left equals the sum of the factors on the right! This does happen occasionally. In particular it has happened for one odd numbered Teaser in the past decade.

What is the number of that Teaser?

170 | FLIPPING AGES!

At a recent family party the ages of the ten people present formed an arithmetic progression (ie, the difference between each person's age in years and the age of the next oldest person was always the same). My age was like my wife's but with the digits in reverse order. Likewise, my sister's age was the reverse of my mother's, my son's age was the reverse of my grandson's, and my daughter's age was the reverse of my granddaughter's. Furthermore, the product of my brother's age and the age of his son equalled the year of birth of one of the people at the party.

What were the ages of my brother, my sister and myself (in that order)?

171 | PROBLEMBLEM

Every Sunday customers in our pub try to solve the current Teaser, so Pat the barman has designed an appropriate pub logo. This is a small rectangular flag divided into three smaller rectangles, coloured red, green and blue. Their sides are all whole numbers of centimetres, the area of the green rectangle is twice the area of the red rectangle, and the perimeter of the red rectangle is twice the perimeter of the green rectangle. Furthermore, the area of the flag is a three-figure number of square centimetres, the same as the area of the blue rectangle but with the three digits in the reverse order.

What is the area of the flag?

172 | BASTILLE DAY

Three towns in France issue their own car licence plates. Argentan's numbers consist of the digits 1 to 9 in some order, Beaurepaire's consist of the digits 1 to 8 in some order, and Corbigny's consist of the digits 1 to 7 in some order. All three towns only use numbers divisible by 11 and all such possible plates have been issued. Tomorrow is Bastille Day and, in order to minimise chaos, only cars with licence numbers divisible by 5 will be allowed in. In the form "one in ...",

(a) **What proportion of Argentan's cars will be allowed into Argentan?**
(b) **What proportion of Beaurepaire's cars will be allowed into Beaurepaire?**
(c) **What proportion of Corbigny's cars will be allowed into Corbigny?**

173 | CELEBRITY GARDENERS

There were six celebrity gardeners at the local show: Beardshaw, Dimmock, Flowerdew, Greenwood, Klein and Thrower. Each of them was exhibiting two different shrubs from the following list: Abutilon, Azalea, Berberis, Cornus, Euonymus, Hibiscus, Holly, Magnolia, Potentilla, Spiraeas, Weigela and Wisteria. On each of six tables there was a card containing the gardener's surname and the name of the two shrubs. For any two of the words on the same card, there were exactly two letters of the alphabet that occurred (perhaps more than once) in both.

(a) Which two shrubs did Beardshaw exhibit?
(b) Who exhibited the Abutilon?

174 | IN THE PUB

George and Martha are in the pub. He has ordered a glass of lager for her and some ale for himself. He has written three numbers in increasing order, none involving a zero, then consistently replaced digits by letters to give

DRANK GLASS LAGER.

George explains this to Martha and tells her that the third number is in fact the sum of the previous two. From this information she is able to work out the value of each letter.

What's the number of George's ALE?

175 | SPINNERS

I have three "spinners" each with five evenly-spaced numbers on. All the numbers on the spinners are positive whole numbers and the sum of the five numbers on each spinner is the same. The first spinner has numbers 3, 3, 3, 3, 3 and the second has numbers 2, 2, 2, 3, 6. My friend chooses a spinner and I choose one of the remaining pair. Then we each spin to choose a number: sometimes it's a draw, otherwise the higher number wins – and we repeat this many times. He gets very annoyed because (whichever spinner he chooses) I am always more likely to win.

What are the five numbers on the third spinner?

176 | GOOD TIMES

The Good Times bus company has a route from Timesboro to Teaseboro which passes through six other towns on its direct route. The distances from each town to the next are different whole numbers of miles, the largest of the seven distances being six miles more than the shortest. I worked out the average of those seven distances and my friend worked out the average distance between any two of the eight towns. Our answers used the same two digits but in reverse orders.

How far is it from Timesboro to Teaseboro?

In his magic cave Ali Baba discovers a string necklace comprising of sixty various gold beads that he decides to melt down and turn into coins. Each bead is spherical but with a cylindrical hole symmetrically cut through it, and in each bead the hole is one centimetre long. Ali Baba melts down all these beads and makes the gold into circular coins of two centimetres diameter and half a centimetre thickness.

How many coins does he obtain?

The 51 states of America have two-letter abbreviations, namely

AK AL AR AZ CA CO CT DC DE
FL GA HI IA ID IL IN KS KY LA
MA MD ME MI MN MO MS
MT NC ND NE NH NI NM NV
NY OH OK OR PA RI SC SD
TN TX UT VA VT WA WI WV WY

I have written as many different letters as possible around a circle such that, if you look at any adjacent pair of letters, then clockwise they are one of the above abbreviations.

One of my letters is R. Starting at R, list the letters clockwise.

179 | STERLING ACHIEVEMENT

Some people were asked to donate one pound each to charity. They all used only "silver" coins (5p, 10p, 20p, 50p) to make up their pound and no two of them used the same combination. I also donated a pound using silver coins, but it was inevitable that my combination of coins would be a repeat of one already used. When all the coins were sorted into the different denominations we found that we had a whole number of pounds in each denomination.

Which coins did I use to make my pound?

180 | MEGAN'S HOUSE

Megan's class has been learning about primes and squares. So their teacher challenged each of them to use some of the digits 1 to 9 to write down three two-figure squares and a two-figure prime without using any digit more than once. They all succeeded in doing this and in Megan's case the left-over digit was her house number. If you knew her house number it would be possible to work out the four two-figure numbers that she wrote down.

What were they?

181 | NEW TRICKS

I am currently experiencing the joy of seeing Liam, my first grandchild, developing lots of new tricks. With this in mind I have written down four odd numbers and then consistently replaced different digits by different letters to give

SIT CRAWL WALK TALK.

In fact the largest of these numbers is the sum of the other three.

Please send in your numerical TRICK.

182 | VERY SIMILAR TRIANGLES

Two triangles are "similar" if they are the same shape (but not necessarily the same size). I cut out two similar (but not identical) triangles from a piece of A4 paper, all the sides of the triangles being whole numbers of centimetres in length. In fact the two triangles were extra similar in the sense that two of the sides of the first triangle were the same lengths as two of the sides of the second triangle.

What were the lengths of the sides of the smaller triangle?

183 | GROUP OF DEATH

In a Champions League group four teams play each other twice with the top two qualifying for the next stage. If two teams are level on points then their head-to-head results determine their position. In one group, after five games each, the teams in order were A, B, C, D, each with a different number of points. At that stage team A was bound to qualify and teams B and C could still top the group. But the final two games were draws. All games had different results, no game had more than five goals, and team D had the same number of "goals for" as "goals against". Team A scored fewer goals than any other team whilst D scored more than any other.

What were the results of the two games BvC?

184 | COLOUR CODED

Five lads cycle together daily. Each has a different number of helmets, less than a dozen, and none of them has two of the same colour. Each wears in daily rotation all the cycle helmets he possesses. On September 1 Alan wore mauve, Bill and Charles wore red, Dave orange and Eric green. On 11th two were red, one green, one mauve and one white. On 19th Dave's was orange, Eric's green and the others red. Eric wore orange on the 22nd and white on the 23rd. On October 1 all five wore the same as on September 1.

In alphabetical order of the riders, what were the helmet colours on the 11th?

185 | MILLENNIAL SQUARES

I asked my class to take some pieces of card and to write a digit on each side of each card so that any year of this millennium that is a perfect square could be spelt out using four of the cards. One clever pupil achieved this with the smallest number of cards possible. He also pointed out that with his cards he could go way beyond the end of this millennium – in fact he had designed them so that, when trying to make this list of consecutive square years yet to come, he was able to go as far as possible with that number of cards.

What is the last square he could make?

186 | ROUND THE GARDEN

George and Martha's garden is a quadrilateral with each of its sides a whole number of metres in length. When Martha wrote down the lengths of the four sides together with the length of the perimeter, she noticed that overall these five numbers used each of the ten digits exactly once. Also, the lengths of the three longest sides of the garden were prime numbers. She told George, but he was unable to work out all the lengths until she also told him that no two of the four lengths differed by a prime number.

What are the four lengths?

187 | BY THE DOZEN

I have written down a five-figure number that uses five different non-zero digits, and then I have replaced the digits by letters to give DOZEN. Within this it is possible to see various numbers, namely

D, O, Z, E, N

DO, OZ, ZE, EN

DOZ, OZE, ZEN

DOZE, OZEN

DOZEN

Surprisingly, the original number is divisible by a dozen or more of these.

What is the number DOZEN?

188 | POWERFUL NUMBERS

George and Martha's five daughters have been investigating large powers of numbers. Each daughter told the parents that they had worked out a number which was equal to its last digit raised to an exact power. When Andrea announced her power, the parents had a 50% chance of guessing the last digit. When Bertha announced a larger power, the same applied. When Caroline announced her power, the parents had only a 25% chance of guessing the last digit; with Dorothy's larger power, the same applied. When Elizabeth announced hers, the parents only had a 12.5% chance of guessing the last digit. I can tell you that the five powers were consecutive two-digit numbers adding up to a perfect square.

What, in alphabetical order of the daughters, were the five powers?

189 | SNAKING UP

A "Snakes and Ladders" board consists of a ten-by-ten grid of squares. In the first row (at the bottom) the squares are numbered from 1 to 10 from left to right, in the second row the squares are numbered 11 to 20 from right to left, in the third they are numbered 21 to 30 from left to right again, and so on.

An ant started on square 1, moved to square 2 and then, always moving to an adjacent square to the right or up, it finished in the top-right corner of the board. I have added up the total of the numbers on the squares it used and, appropriately, the total is a perfect square. In fact it is the square of one of the numbers the ant visited – the ant passed straight through it without turning.

What was that total of the numbers used?

190 | COINS OF OBSCURA

George and Martha were on holiday in Obscura where coins are circular with radii equal to their whole number value in scurats. They had coffees and George took out some loose change. He placed one 10 scurat and two 15 scurat coins on the table so that each coin touched the other two and a smaller coin between them that touched all three. He then placed a fifth coin on top of these four which exactly covered them and the total value of the five coins was exactly enough to pay the bill.

How much was the coffee bill?

191 | ON A ROLL

My son saw the "average sheets per roll" printed on our new pack of four toilet rolls. Curious, he counted each roll's total sheets by overlaying lines of ten sheets to and fro, tallying layers, and then adding any extra sheets. These totals were four consecutive three-figure numbers, including the printed "average sheets per roll". For each total, he noticed that there was the same single-figure number of choices for a number of sheets per layer, requiring at least two layers, which would leave no extra sheets. Recounting the toilet roll with the "average sheets per roll", he used a two-figure number of sheets per layer and tallied a two-figure number of such layers, with no extras.

What was "the average sheets per roll"?

192 | GOLF BALLS

Three friends play 18 holes of golf together every week. They have a large supply of golf balls numbered from 1 to 5. At the beginning of a year, they each start playing with a new ball, but all three ball numbers are different. Alf uses the same ball number for exactly 21 holes before changing to the next higher number (or to number 1 if he was using a number 5). He continues to use each number for exactly 21 holes. The same applies to Bert, except that he changes his ball number in the same way after every 22 holes, and to Charlie who changes his ball number after every 23 holes.

Last year, there was just one occasion when they all used the same ball number on a hole.

What was the number of the hole on that occasion?

193 | MY PIN NUMBERS

Using all but one of the digits 0 to 9, and systematically replacing them by a letter, my pin numbers become ONE, TWO, FIVE and SEVEN. These numbers are such that

- ## ONE IS ODD
- ## TWO IS EVEN
- ## FIVE IS ODD AND DIVISIBLE BY 5
- ## SEVEN IS DIVISIBLE BY 7
- ## ONE + TWO + TWO = FIVE

If I told you which digit was not used, you should be able to work out my pin numbers.

What pin number is represented by SEVEN?

194 | ODD SOCKS

I had a drawer containing some black socks and some white socks. If I drew out two socks at random the chance of getting a black pair was 1 in ...

After many washes all the socks looked grey. So I added some red socks to the drawer. Then if I drew out two at random the chance of getting a grey pair was 1 in ...

After many washes all the socks looked pink. So I added some green socks to the drawer. Then if I drew out two the chance of getting a pink pair was 1 in ...

After many washes all the socks looked brown. So I have now added some yellow socks to the drawer giving me a total of fewer than fifty socks. Now if I draw out two the chance of getting a brown pair is 1 in ...

The gaps above consist of four different prime numbers.

If I draw out two socks at random, what is the chance of getting a yellow pair?

195 | UNFORTUNATE 57

In the early days of the internet, I used a secret shorthand for my important passwords: Bank=1/7, Credit Card=2/7, ISP=3/7, etc. Like all fractions, the decimal expansions

$$1/7 = 0.142857142857142\ldots, 2/7 = 0.285714285714285\ldots$$

eventually repeat themselves, in this case in sequences of six digits. In each case, my password was the set of digits that repeat ('Unfortunate 57" is a mnemonic for 142857). As password requirements became stricter, I changed my system to base 11, using an X for the extra digit for 'ten'; so for instance in base 11

234 ($=1\times11^2+10\times11+1\times3$) is $1X3$ and $1/2=0.5555\ldots = 5/11+5/(11^2)+5/(11^3)+\ldots$

In the sequence 1/2, 1/3,…, what is the first password of length greater than six that my base-11 system produces?

196 | TRIANGULATION

Liam plans to make a set of dominoes. They will be triangular, and one face of each domino will have a number at each corner. The numbers run from 0 up to a maximum digit (9 or less), and the set is to include all possible distinguishable dominoes.

With the maximum digit he has chosen the set would contain a triangular number of dominoes. [A triangular number is one where that number of balls can fit snugly in an equilateral triangle, for example the 15 red balls on a snooker table.]

How many dominoes will he need to make?

197 | SUNDAY TEASER

I wrote down two three-figure numbers and worked out their product by long multiplication. Systematically replacing digits by letters, my workings became

```
        N   T   S
        E   D   S
    ─────────────────
        D   U   R   S
    Y   R   D
A   N   D   U
─────────────────────
A   R   R   A   T   S
```

I then wrote down two numbers which were represented by SUNDAY TEASER.

What were these two numbers?

198 | GOOD ARRAZ, BAZ!

Baz's three darts hit the board, scoring different numbers from 1 to 20. "Curious numbers," said Kaz. Baz looked puzzled. Kaz explained that the first dart's score to the power of the second dart's score is a value that contains each numeral 0 to 9 at least once and has the third dart's score number of digits. Baz only saw that the third dart's score was the difference between the other two darts' scores. Kaz wrote the full value on a beer mat. Then Baz put his glass on it and covered most of the value, leaving just the first dart's score showing to the right and the second dart's score to the left.

What did each dart score in the order thrown?

199 | A PALINDROME

In this Teaser, a jig* is defined as an outwards move to an adjacent empty square, either horizontally, upwards or downwards, the letter * being inserted in all such squares. Begin with the letter W on a regular grid of empty squares.

From the W, jigO. From every O, jigN. From every N, jigD, and so on until the central diagonal reads

SELIM'S TIRED, NO WONDER, IT'S MILES.

Looking at your grid of letters, in how many ways can you trace the palindrome above?

[You can start at any S, move to adjacent letters till you reach the W and then on to any S (including the one you started at). You may move up and down, left and right.]

200 | MULTICOLOURED

George and Martha are selling wall-paper of various colours. By replacing letters with positive digits, they have devised the following multiplication problem:

RED

X GREY

= YELLOW

The N in GREEN is the remaining positive digit. The red wall-paper is sold only in squares, which is appropriate since RED is a perfect square.

What is the value of GREEN?

SOLUTIONS

TEASER 1
33 degrees per minute

TEASER 2
24961353

TEASER 3
Danny Roth and Victor Bryant

TEASER 4
125928

TEASER 5
9841788

TEASER 6
226 and 135

TEASER 7
7

TEASER 8
7560

TEASER 9
£54

TEASER 10
£1019

TEASER 11
74 poles

TEASER 12
33

TEASER 13
144

TEASER 14
735 sq cm

TEASER 15
1535

TEASER 16
18 and 72

TEASER 17
180 years

TEASER 18
44 metres

TEASER 19
(a) 35 runs; (b) 25 runs

TEASER 20
7146 or 7543, with apologies for non-uniqueness

TEASER 21
2

TEASER 22
168

TEASER 23
Roddy 25 cards

TEASER 24
672

TEASER 25
105

TEASER 26
£74,536

TEASER 27
204in or 17ft

TEASER 28
4 and 536870912

TEASER 29
85 metres

TEASER 30
5-5

TEASER 31
112

TEASER 32
467

TEASER 33
76398245

TEASER 34
230769

TEASER 35
5

TEASER 36
11

TEASER 37
263, 841 and 907

TEASER 38
12589

TEASER 39
489051765

TEASER 40
Alan Collins; dachshund

TEASER 41
13

TEASER 42
919 and 787

TEASER 43
16 and 61

TEASER 44
17

TEASER 45
LXII

TEASER 46
143968275

TEASER 47
(a) 16 (b) 2p, 10p and £2

TEASER 48
3, 4 and 5

TEASER 49
5670000

TEASER 50
24 boys and 8 girls

TEASER 51
Clara, Debbie and Erica

TEASER 52
89115

TEASER 53
5476

TEASER 54
34

TEASER 55
121, 125, 216, 256 and 512

TEASER 56
73

TEASER 57

12.30

TEASER 58

62

TEASER 59

1 2 5

3 4 6

7 8 9

TEASER 60

Amelia, 93

TEASER 61

(a) 2100010006 (b) 100000000

TEASER 62

18383838383838383838382

TEASER 63

5%

TEASER 64

217 metres

TEASER 65

1, 4 and 9

TEASER 66

98, 86, 68 and 41

TEASER 67

361 and 625

TEASER 68

2435160

TEASER 69

01235

TEASER 70
132

TEASER 71
158453

TEASER 72
157326849

TEASER 73
23 49 50 67 81 and 46970 83521

TEASER 74
4539

TEASER 75
Smith, 2nd in the queue

TEASER 76
6, 11 and 24

TEASER 77
Charlie's team finished second and the other two teams drew.

TEASER 78
11

TEASER 79
15, 21 and 35

TEASER 80
7

TEASER 81
Pisces, Virgo, Leo, Virgo and Leo

TEASER 82
387

TEASER 83
ANSWER (!)

TEASER 84
2nd

TEASER 85
18, 26 and 37

TEASER 86
4cm and 8cm

TEASER 87
5 by 15 by 15

TEASER 88
972364815

TEASER 89
4232

TEASER 90
1, 4, 1 and 2

TEASER 91
10552

TEASER 92
11.43

TEASER 93
2143

TEASER 94
2727

TEASER 95
798

TEASER 96
5, 6 and 7cm

TEASER 97
12

TEASER 98
25, age 73

TEASER 99
30 metres

TEASER 100
March, June, May, July, October

TEASER 101
2, 7, 8 and 13

TEASER 102
(a) Devil's Um and The Gypsy. (b) Blue Toothed and Slippery Jack.

TEASER 103
£108

TEASER 104
6/37, 14/37 and 17/37

TEASER 105
278521

TEASER 106
1, 2, 4, 5, 7, 8, 10, 12, 14, 17, 20

TEASER 107
2, 3, 80, 491, 576

TEASER 108
531713717

TEASER 109
Beth, 819

TEASER 110
612

TEASER 111
45

TEASER 112
Mike Smith, Ivan Perkins, Ken Jones, Nigel Brown
and John Robinson

TEASER 113
17, 13, 11, 19, 7, 3

TEASER 114
5901

TEASER 115
Tuesday

TEASER 116
Glen, Friday and June

TEASER 117
3024

TEASER 118
124421124

TEASER 119
1x2, 2x3, 8x9

TEASER 120
6 furlong and 29 furlong

TEASER 121
2, 4,5 and 6

TEASER 122
30 feet

TEASER 123
70cm

TEASER 124
983 and 821

TEASER 125
100954

TEASER 126
1, 3, 6, 7, 9, 10, 12, 15, 21 and 28

TEASER 127
523

TEASER 128
105625

TEASER 129
20

TEASER 130
2301

TEASER 131
6739

TEASER 132
6, 8 and 10 metres

TEASER 133
TIME=5269

TEASER 134
7

TEASER 135
13, 69, 20, 84 and 57

TEASER 136
A:7 and B:2

TEASER 137
112

TEASER 138
1, 5, 7, 8, 9, 10, 11 and 13 metres

TEASER 139
94267830

TEASER 140
22 and 24 minutes

TEASER 141
495, 702, 816

TEASER 142
(a) 21 (b) 42 (c) 29 and 34

TEASER 143
(a) 51; (b) 11, 23, 31 and 41

TEASER 144
37

TEASER 145
(a) 2-by-35 (b) 2016

TEASER 146
27

TEASER 147
6

TEASER 148
16561 and 78487

TEASER 149
58225185

TEASER 150
523

TEASER 151
Dalton, Cazenove, Connery, Lazenby, Craig, Moore, Dench, Brosnan

TEASER 152
19, 64, 76 and 91

TEASER 153
296143

TEASER 154
Fred, Big Gun

TEASER 155
£14.56

TEASER 156
135 cm^2

TEASER 157
200546

TEASER 158
16

TEASER 159
5 and 401

TEASER 160
147 and 3285

TEASER 161
20, 6 and 4

TEASER 162
Mr Cook and Mrs Cook

TEASER 163
1, 4, 5, 7, 9 and 14cm

TEASER 164
1820, 1836, 1849, 1859, 1870, 1890, 1892 and 1904

TEASER 165
8654310

TEASER 166
8237

TEASER 167
14

TEASER 168
LOW = 592

TEASER 169
2401

TEASER 170
60, 69 AND 78

TEASER 171
231 CM^2

TEASER 172
(a) 1 in 11; (b) 1 in 8; (c) 1 in 8

TEASER 173
(A) CORNUS AND POTENTILLA; (B) GREENWOOD

TEASER 174
392

TEASER 175
1, 1, 4, 4, 5

TEASER 176
98 miles

TEASER 177
20 coins

TEASER 178
R I L A K S D C T N M O

TEASER 179
two 20p, three 10p and six 5p

TEASER 180
16, 25, 49 and 83

TEASER 181
TRICK = 50619

TEASER 182
8cm, 12cm and 18cm

TEASER 183
4-1 and 1-1

TEASER 184
mauve, red, red, green, white

TEASER 185
3721

TEASER 186
4, 53, 61 and 89 metres

TEASER 187
31248

TEASER 188
44, 46, 43, 47, 45

TEASER 189
676

TEASER 190
72 scurats

TEASER 191
242

TEASER 192
10

TEASER 193
SEVEN = 95452

TEASER 194
1 in 6

TEASER 195
093425X17685

TEASER 196
45

TEASER 197
684219 and 731635

TEASER 198
5, 19 and 14

TEASER 199
1073479696

TEASER 200
21668

PUZZLE SETTERS

TEASER 1
Danny Roth

TEASER 2
Angela Newing

TEASER 3
Victor Bryant

TEASER 4
Andrew Skidmore

TEASER 5
H Bradley and C Higgins

TEASER 6
Victor Bryant

TEASER 7
Andrew Skidmore

TEASER 8
DJT Hogg

TEASER 9
Danny Roth

TEASER 10
H Bradley and C Higgins

TEASER 11
Graham Smithers

TEASER 12
Victor Bryant

TEASER 13
Tom Wills-Sandford

TEASER 14
H Bradley and C Higgins

TEASER 15
Graham Smithers

TEASER 16
Robin Nayler

TEASER 17
Danny Roth

TEASER 18
Brian Gladman

TEASER 19
Graham Smithers

TEASER 20
Victor Bryant

TEASER 21
Andrew Skidmore

TEASER 22
Danny Roth

TEASER 23
Graham Smithers

TEASER 24
Bob Walker

TEASER 25
Andrew Skidmore

TEASER 26
Graham Smithers

TEASER 27
H Bradley and C Higgins

TEASER 56
Adrian Somerfield

TEASER 57
H Bradley and C Higgins

TEASER 58
Robin Nayler

TEASER 59
Graham Smith

TEASER 60
Danny Roth

TEASER 61
Simon Massarella

TEASER 62
Victor Bryant

TEASER 63
Andrew Skidmore

TEASER 64
Rex Lanham

TEASER 65
Nick MacKinnon

TEASER 66
H Bradley and C Higgins

TEASER 67
Victor Bryant

TEASER 68
Graham Smithers

TEASER 69
John Owen

TEASER 84
Angela Newing

TEASER 85
H Bradley and C Higgins

TEASER 86
Des MacHale

TEASER 87
Andrew Skidmore

TEASER 88
Graham Smithers

TEASER 89
Peter G Chamberlain

TEASER 90
Angela Newing

TEASER 91
Andrew Skidmore

TEASER 92
Robin Nayler

TEASER 93
Victor Bryant

TEASER 94
Nick MacKinnon

TEASER 95
Graham Smithers

TEASER 96
H Bradley and C Higgins

TEASER 97
Peter G Chamberlain

TEASER 98
Danny Roth

TEASER 99
Des MacHale

TEASER 100
Robin Nayler

TEASER 101
Victor Bryant

TEASER 102
Graham Smithers

TEASER 103
H Bradley and C Higgins

TEASER 104
Nick MacKinnon

TEASER 105
Andrew Skidmore

TEASER 106
Danny Roth

TEASER 107
Victor Bryant

TEASER 108
Peter G Chamberlain

TEASER 109
Graham Smithers

TEASER 110
Robin Nayler

TEASER 111
H Bradley and C Higgins

TEASER 112
Angela Newing

TEASER 113
Peter G Chamberlain

TEASER 114
Andrew Skidmore

TEASER 115
John Owen

TEASER 116
Victor Bryant

TEASER 117
Graham Smithers

TEASER 118
Peter G Chamberlain

TEASER 119
Des MacHale

TEASER 120
Rex Lanham

TEASER 121
Robin Nayler

TEASER 122
H Bradley and C Higgins

TEASER 123
Andrew Skidmore

TEASER 124
Peter G Chamberlain

TEASER 125
Victor Bryant

TEASER 126
Danny Roth

TEASER 127
Graham Smithers

TEASER 128
Peter G Chamberlain

TEASER 129
Des MacHale

TEASER 130
John Owen

TEASER 131
Angela Newing

TEASER 132
Alan Bergson

TEASER 133
Victor Bryant

TEASER 134
H Bradley and C Higgins

TEASER 135
Ian Duff

TEASER 136
Peter G Chamberlain

TEASER 137
Graham Smithers

TEASER 138
Danny Roth

TEASER 139
Andrew Skidmore

TEASER 140
Robin Nayler

TEASER 141
Victor Bryant

TEASER 142
Nick MacKinnon

TEASER 143
Peter G Chamberlain

TEASER 144
Victor Bryant

TEASER 145
Graham Smithers

TEASER 146
Danny Roth

TEASER 147
Ian Duff

TEASER 148
David Buontempo

TEASER 149
Peter G Chamberlain

TEASER 150
Victor Bryant

TEASER 151
Graham Smithers

TEASER 152
Robin Nayler

TEASER 153
Danny Roth

TEASER 154
Angela Newing

TEASER 155
H Bradley and C Higgins

TEASER 156
Victor Bryant

TEASER 157
Andrew Skidmore

TEASER 158
Des MacHale

TEASER 159
Graham Smithers

TEASER 160
Victor Bryant

TEASER 161
Robin Nayler

TEASER 162
Angela Newing

TEASER 163
H Bradley and C Higgins

TEASER 164
Nick MacKinnon

TEASER 165
Danny Roth

TEASER 166
Andrew Skidmore

TEASER 167
Graham Smithers

TEASER 168
Peter G Chamberlain

TEASER 169
Victor Bryant

TEASER 170
Robin Nayler

TEASER 171
H Bradley and C Higgins

TEASER 172
Tom Wills-Sandford

TEASER 173
Graham Smithers

TEASER 174
Danny Roth

TEASER 175
Mike Fletcher

TEASER 176
H Bradley and C Higgins

TEASER 177
Nick Jones

TEASER 178
Victor Bryant

TEASER 179
Robin Nayler

TEASER 180
Graham Smithers

TEASER 181
Andrew Skidmore

TEASER 182
Michael Fletcher

TEASER 183
Nick MacKinnon

TEASER 184
Angela Newing

TEASER 185
Adrian Somerfield

TEASER 186
Danny Roth

TEASER 187
Victor Bryant

TEASER 188
Danny Roth

TEASER 189
Victor Bryant

TEASER 190
Peter Good

TEASER 191
Stephen Hogg

TEASER 192
John Owen

TEASER 193
Graham Smithers

TEASER 194
Victor Bryant

TEASER 195
Susan Bricket

TEASER 196
Andrew Skidmore

TEASER 197
Graham Smithers

TEASER 198
Stephen Hogg

TEASER 199
Graham Smithers

TEASER 200
Danny Roth